Chapter 4

Bearing Capacity Failure: Transcona Grain Elevator, Canada

4.1 Case Description

In September 1913, the Canadian Pacific Railway Company completed construction of a million-bushel (about 36,400 m^3) grain elevator at North Transcona, 11 km north-east of Winnipeg, Canada. The elevator was one of the most important structures and one of the largest gravity railroad yards in the world, which covered several square miles and was built on partly farmed, relatively flat prairie land. The purpose of the elevator was to provide relief for the Winnipeg Yards during the months of peak grain-shipment.

The structure consisted of a reinforced-concrete work-house, and an adjoining bin-house, which contained five rows of 13 bins, each 28 m in height and 4.4 m in diameter. The bins were based on a concrete structure containing belt conveyors supported by a reinforced-concrete shallow raft foundation.

After the structure was completed, the filling was begun and grain was distributed uniformly between the bins. On October 18, 1913, after the elevator was loaded to 87.5% of its capacity, settlement of the bin-house was noted. Within an hour, the settlement had increased uniformly to about 30 cm following by a tilt towards the west (Fig. 4.1a,b), which continued for almost 24 hours until it reached an inclination of almost 27 degrees (Allaire, 1916).

(a) (b)

Figure 4.1 The collapse of the Transcona Grain Elevator: (a) view looking southwest (Engineering News, 1913); (b) view looking northeast (White, 1953: © 1953 Thomas Telford Limited. Used with permission).

Many years later, when soil mechanics had provided the basis for computing the ultimate bearing-capacity of soils, it was realized that the Transcona failure

A.M. Puzrin • E.E. Alonso • N.M. Pinyol

Geomechanics of Failures

Springer

A.M. Puzrin
Institut für Geotechnik
ETH Zürich
Switzerland
alexander.puzrin@igt.baug.ethz.ch

E.E. Alonso
Civil Engineering School
UPC
Barcelona, Spain
eduardo.alonso@upc.es

N.M. Pinyol
CIMNE
UPC
Barcelona, Spain
nuria.pinyol@upc.edu

ISBN 978-94-007-9397-2   ISBN 978-90-481-3531-8 (eBook)
DOI 10.1007/978-90-481-3531-8
Springer Dordrecht Heidelberg London New York

Softcover re-print of the Hardcover 1st edition 2010

Printed on acid-free paper

Springer is part of Springer Science+Business Media (www.springer.com)

Table of Contents

Preface

The main goal of this introductory text is to demonstrate how basic concepts in Soil Mechanics can be used as a "forensic" tool in the investigation of geotechnical failures. This, in turn, provides a good opportunity to show how to use available procedures in the formulation of useful simple models. Geotechnical failure is understood here in a broad sense as the failure of a structure to function properly due to a geotechnical reason.

Some of the geotechnical failures selected are well known for their impact on the geotechnical community. Others are closer to the author's experience. They have been organized into three main topics: Settlement, Bearing Capacity and Excavations. They cover a significant proportion of every day's activity of professional geotechnical engineers. No attempt has been made to create a comprehensive handbook of failures. Instead, the emphasis has been given to creative applications of simple mechanical concepts and well known principles and solutions of Soil Mechanics. The book shows how much can be learned from relatively simple approaches. Despite this emphasis on simplicity, the book provides a deep insight into the cases analyzed. A non-negligible number of new analytical closed-form solutions have also been found. Their derivation can be followed in detail.

In all the cases described an effort was made to provide a detailed and step by step description of the hypothesis introduced and of the analysis performed. Each of the eight chapters of the book addresses a certain type of failure, illustrated by a case history. The chapters have a common structure which is essentially the following:

1. Case description
2. Relevant theory
3. Mechanical analysis
4. Mitigation measures
5. Lessons learned

The chapters are self-contained. They provide a review of Soil Mechanics principles and methods required to understand and explain the failure described. In some cases the analysis offered provides a non-conventional application of basic principles.

An often asked question regarding failures is how to avoid them. This is of great practical interest and a section of each chapter is dedicated to provide a few solutions. The chapters are completed with a summary of lessons learned from the failure and its analysis. Some chapters also include a short account on advanced topics to help the interested readers to go beyond the approaches used in the book.

Readers are expected to be familiar with the basic concepts of Soil Mechanics and Foundation Engineering. The target audience is undergraduate and graduate students, faculty and practicing professionals in the fields of Civil and Geotechnical Engineering.

This text book benefits from the experience accumulated in teaching undergraduate and graduate courses in Geotechnical and Forensic Engineering at the ETH Zurich. The authors would like to express their gratitude to Carlo Rabaiotti, Michael Iten, Stefan Annen, Erich Saurer and Andreas Schmid, all of the ETH Zurich, for their contribution to various chapters of this book. Markus Schwager, Martina Baertsch, Simon Sauter and Pascal Minder helped to prepare the manuscript. The support provided by Javier Uzcanga (Port Authority of Barcelona), Ricardo Madrid and Dani Tarragó, both from the UPC, during the preparation of Chapter 5 is also greatly acknowledged. The authors express their gratitude to Raúl Giménez and María del Mar Obrador for their dedication to the drafting of figures and formatting of the text.

A.M. Puzrin, E.E. Alonso and N.M. Pinyol
Zürich, Barcelona, December 2009

PART I

SETTLEMENTS

Chapter 1

Interaction between Neighbouring Structures: Mexico City Metropolitan Cathedral, Mexico

TABLE OF CONTENTS

A.M. Puzrin et al., *Geomechanics of Failures*, DOI 10.1007/978-90-481-3531-8_1,

Chapter 1
Interaction between Neighbouring Structures: Mexico City Metropolitan Cathedral, Mexico

1.1 Case Description

The Metropolitan Cathedral in Mexico City (Fig. 1.1) is one of the most spectacular and important architectural monuments in the Americas. It was built in the sixteenth century on extremely soft lacustrine clays over the remains of ancient Tenochtitlan – the lake capital of the Aztec empire (Fig. 1.2).

Alarmingly large differential settlements of the Cathedral and the adjacent Metropolitan El Sagrario parish church have accumulated over the centuries and presently threaten the stability of the structures.

Figure 1.1 The Metropolitan Cathedral and El Sagrario parish church in Mexico City (© David Alayeto).

1.1.1 Construction

Construction of the Metropolitan Cathedral was started around 1560. Its shallow foundation rests on the artificial fill layer placed over alternating thick layers of very soft saturated lacustrine clay and sand (Fig. 1.3). The foundation is built on a 3.5 m thick basement made from a grid of masonry and beams, placed over a 2 m thick layer of rock masonry which rests on a 30 cm thick grid of short stakes. This

Figure 1.2 Ancient Tenochtitlan (from the mural painting at the National Museum of Anthropology, Mexico City ©. Painted in 1930 by Dr. Atl).

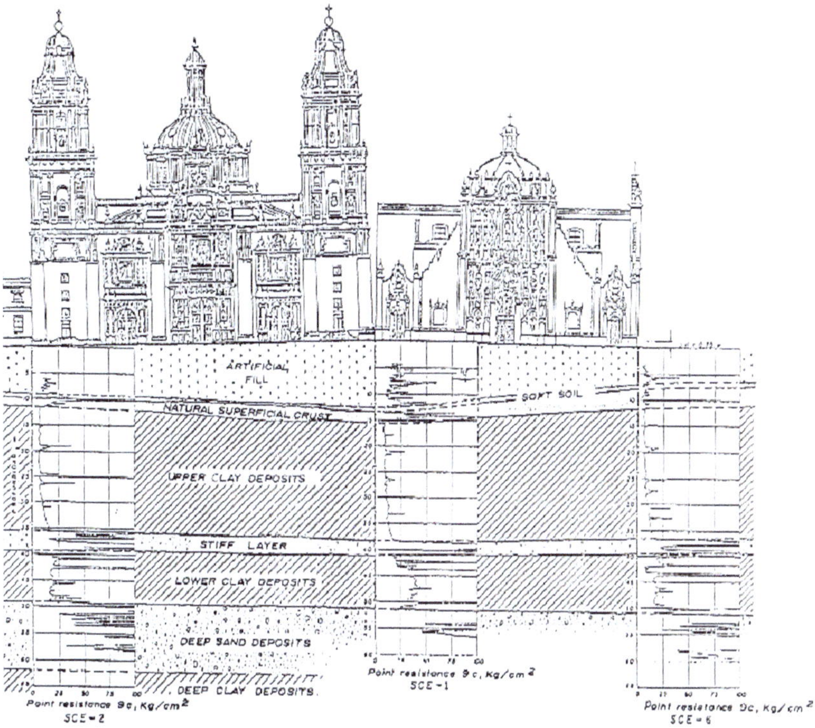

Figure 1.3 Cross-section and the soil profile of the Cathedral and El Sagrario (after Guerra, 1992, reprinted with permission from the Vol. 24, No. 1-2, 1992 issue of the *APT Bulletin, The Journal of the Association for Preservation Technology International*).

foundation system was borrowed from existing local pre-Hispanic practice (a sensible thing to do). The building was made of rock masonry. Its footprint is about 60 m wide and 125 m long, has five naves, a central dome, and two 60 m high east (E) and west (W) bell towers (Fig. 1.4). The average pressure transmitted by the structure to the soil is about 166 kPa (after Santoyo and Ovando, 2002).

The Metropolitan Sagrario parish church was built between 1749 and 1768 (i.e. almost two centuries later). Its foundation consists of a 1.2 m thick rock masonry platform resting on a grid of wooden stakes placed over a 30 cm thick bed of mortar. The footprint of the church (Fig. 1.4) is about 47 by 47 m and the average pressure transmitted to the soil is about 132 kPa.

1.1.2 The history of settlements

It is well documented that the cathedral has undergone continuous differential settlements since the earliest stages of its construction. This can be also deduced from the considerable variations in the foundation platform thickness, different lengths of the columns, and wedged masonry layers.

Surveys performed in 1907 revealed differential settlement of more than 1.5 m between the apse and the western tower (Fig. 1.4); in 1972, this settlement reached 2.2 m and exceeded 2.4 m in 1990. The western tower settled 1.25 m more than the eastern. El Sagrario tilted in the opposite direction with the differential settlement of 0.50 m. In April 1989, heavy rainfall caused leakage, revealing severe cracking along the building's southeast-northwest direction.

Differential settlements continued to increase, in spite of a number of underpinning attempts undertaken in the past. Between 1930 and 1940, the existing foundation was reinforced and a huge foundation raft was incorporated into it. The behaviour of the cathedral temporarily improved. In 1972, the cathedral was underpinned with piles provided with a device allowing for the magnitude of the pile load to be controlled. This measure could have been very effective in leveling the cathedral, provided the piles functioned as expected. Unfortunately, due to the difficulties during pile driving, many piles ended up defective or too short, with significantly reduced bearing capacity.

1.1.3 The problem

As often happens in geotechnical engineering, a number of possible factors contributed to the differential settlements. Among these factors is the construction of the deep drainage pipe along the southern facades of the Cathedral and El Sagrario and of the subway tunnel along the eastern side of El Sagrario (Fig. 1.4). While these factors could help to explain the south-eastern tilt of El Sagrario, they fail to account for the western tilt of the Cathedral.

One of the major factors contributing to the settlements in the Mexico City is the regional subsidence as a consequence of the consolidation of the soft clay layers. It results from the over-exploitation of deep aquifers and the rapid drop in the groundwater level (from 3.5 m below the surface in 1972 to 7.4 m in 1990). But if the groundwater level drops uniformly in the entire region, how can this bring about differential settlements?

This "paradox" can be explained by non-homogeneous distribution of layer thickness and compressibility within the soft clay deposits underlying the Cathedral (Fig. 1.3). These variations, however, are only partially due to the natural geological reasons, which account, according to some estimates (Guerra, 1992), for about 20% of the total differential settlement. A much greater effect on this compressibility variation may have the loading history (Fig. 1.5) and interaction between neighbouring structures, which is the main focus of this chapter.

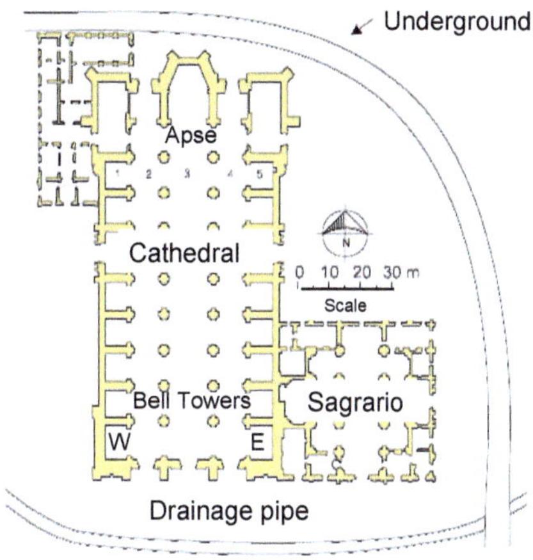

Figure 1.4 Plan of the Cathedral and El Sagrario (after Santoyo and Ovando, 2002).

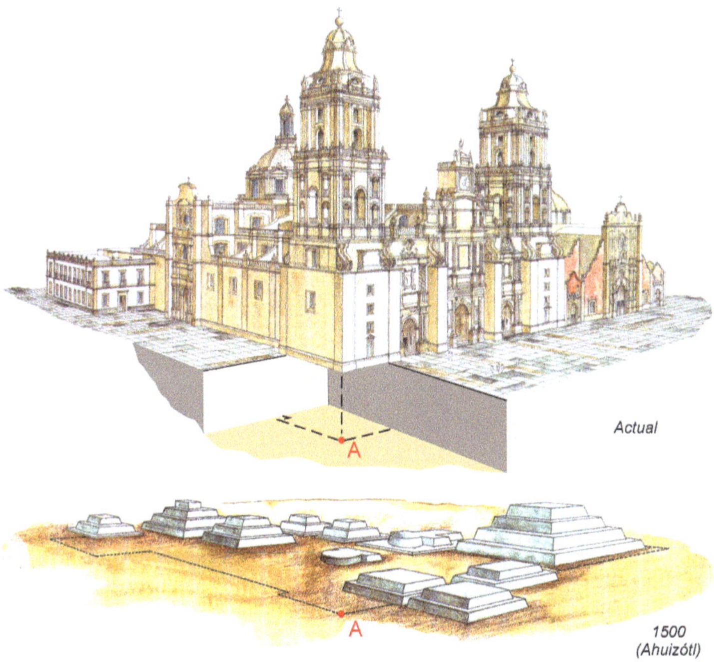

Figure 1.5 Location (relative to the Cathedral and El Sagrario) of the Aztec temples destroyed by Cortés in August 1521 (after Santoyo and Ovando, 2002).

1.1.4 The loading history

Historical records indicate that the Cathedral and El Sagrario were built over the remains of an ancient Aztec temple. The temple was destroyed by Spanish conquistador Hernando Cortés in August 1521, after Tenochtitlan – the capital of the Aztec empire – fell to his army. The largest (and the heaviest) pyramid of the temple (Fig. 1.5) stood within the El Sagrario footprint. Effects of this loading history on differential settlements of the Cathedral and on its interaction with El Sagrario are studied below.

1.2 The Theory

The theory for the analysis of the interaction between the neighbouring structures is best illustrated using the problem of two silos (Fig. 1.6a). We will consider an example of two square 10×10 m^2 silos A and B built at a 2 m distance (Fig. 1.6b). The contact pressure q = 200 kPa. The groundwater level is at the top of the 30 m deep, normally consolidated clay layer underlain by incompressible rock. The interaction between the two silos will be explored in the following three scenarios:

1) Silos A and B are built simultaneously;
2) Silo B is built after Silo A;
3) Silo B is built after Silo A is built and removed.

But first, some necessary theory will be recalled.

Figure 1.6 Two neighbouring silos: (a) a failure case (Bozozuk, 1976; © NRC Canada); (b) example.

1.2.1 Stresses

A total vertical stress increment $\Delta\sigma_z$ at a depth z under a corner of a rectangular area $a \times b$ loaded by a uniform vertical load q is given by

$$\Delta\sigma_z = qJ(a,b,z), \tag{1.1}$$

where J is an influence factor (Lang *et al.*, 2007):

$$J = \frac{1}{2\pi}\left[\arctan\left(\frac{ab}{Rz}\right) + \frac{abz}{R}\left(\frac{1}{a^2+z^2} + \frac{1}{b^2+z^2}\right)\right], \tag{1.2}$$

where $R^2 = a^2 + b^2 + z^2$.

The influence factors for the pressure due to Silo A at points E_1, E_2, E_3, and E_4 on the centerline of the footings of both silos are determined using the superposition of real and fictitious foundations (Fig. 1.7). This is necessary because Equations (1.1) and (1.2) only provide stresses below the corner of the loaded area. Figure 1.7 and the superposition procedure are explained below.

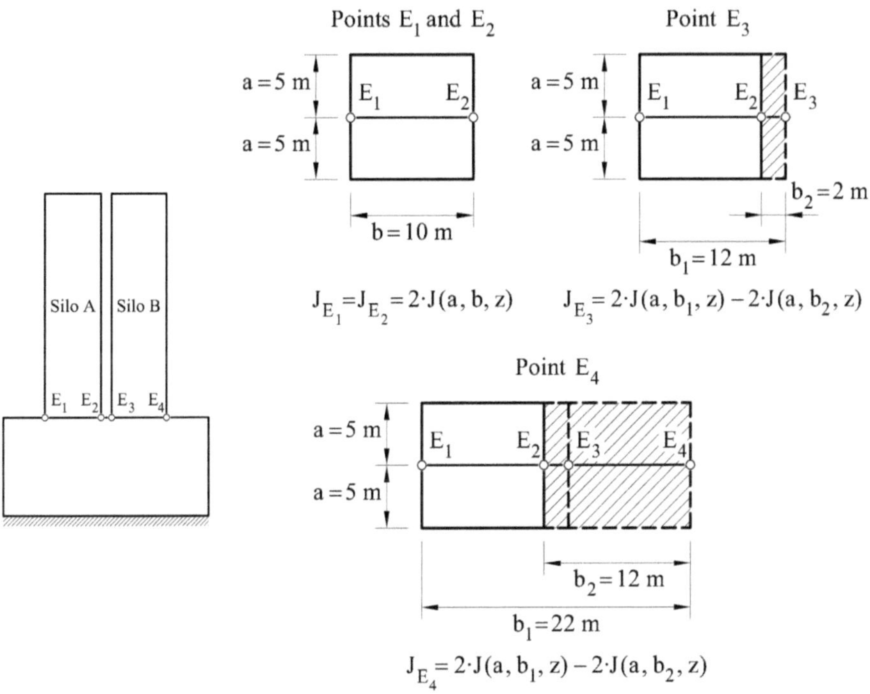

Figure 1.7 Superposition of fictitious foundations: effects of pressure of Silo A on stresses under points E_1–E_4.

The effect of Silo A on the vertical stresses below points E_1 and E_2 is calculated by dividing the 10×10 m^2 silo foundation into two 5×10 m^2 areas having the same uniform load q. Then, points E_1 and E_2 are at the corners of these two areas, and Equations (1.1) and (1.2) can be applied directly. Therefore, the combined influence factor at points E_1 and E_2 is calculated as: $J_{E_1} = J_{E_2} = 2J(5,10,z)$.

The effect of Silo A on the vertical stresses below the point E_3 is calculated by fictitiously extending the real 10×10 m^2 silo foundation by 2 m in order to include the point E_3 into its shorter side. Then, this extended area is divided into two 5×12 m^2 foundations having the same uniform load q. Since point E_3 is at the corner of these two foundations, Equations (1.1) and (1.2) can be applied directly. The next step is to remove the effect of the added fictitious 2×10 m^2 area (hatched area in Fig. 1.7). This area is also divided into two 5×2 m^2 foundations having the same uniform load q as previously. In this way the point E_3 is again at the corner of these two fictitious foundations. The total influence factor is obtained by subtracting the effect of the enlarged and the hatched areas as follows: $J_{E_3} = 2J(5,12,z) - 2J(5,2,z)$.

The effect of Silo A on the vertical stresses below the point E_4 is calculated in a similar way. The real 10×10 m^2 silo foundation is fictitiously extended by 12 m (instead of 2 m) to include the point E_4 into its shorter side, so that the added fictitious part in this case has dimensions 12×10 m^2. The influence coefficient for E_4 is now: $J_{E_4} = 2J(5,22,z) - 2J(5,12,z)$.

1.2.2 Settlements

In general, final settlement of a thin clay layer due to consolidation under the effective stress increment $\Delta\sigma'$ (which, by the end of the consolidation, is equal to the total stress increment $\Delta\sigma' = \Delta\sigma$) is given by Figure 1.8:

$$\Delta\rho = \frac{\Delta H}{1+e_0}\left(\Delta e_1 + \Delta e_2\right) = \frac{\Delta H}{1+e_0}\left(C_e \log\frac{\sigma'_c}{\sigma'_0} + C_c \log\frac{\sigma'_0 + \Delta\sigma}{\sigma'_c}\right). \tag{1.3}$$

where σ'_0 is the in-situ geostatic stress;
e_0 is the void ratio;
σ'_c is the preconsolidation stress:
C_c is the compression index;
C_e is the swelling index;
ΔH is the layer thickness.

For normally consolidated clays ($\sigma'_c = \sigma'_0$),

$$\Delta\rho = \frac{\Delta H}{1+e_0} C_c \log\frac{\sigma'_0 + \Delta\sigma}{\sigma'_0}.$$

For strongly overconsolidated clays (and during any unloading) ($\sigma_c' > \sigma_0' + \Delta\sigma$),

$$\Delta\rho = \frac{\Delta H}{1+e_0} C_e \log \frac{\sigma_0' + \Delta\sigma}{\sigma_0'} .$$

For a thick clay layer, the stresses σ_c', σ_0' and $\Delta\sigma$ cannot be considered constant with the depth of the layer. In this case, the settlement is calculated by dividing the thick layer into a number of sublayers, sufficiently thin to consider stresses σ_c', σ_0' and $\Delta\sigma$ being constant within each of them, and summing their individual settlements: $\Delta\rho = \sum \Delta\rho_i$.

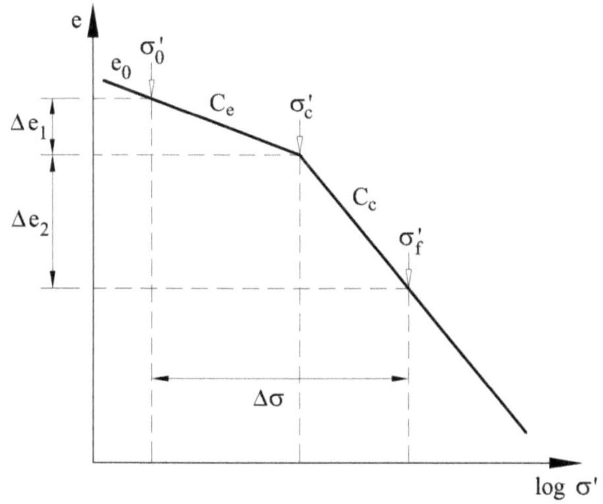

Figure 1.8 Final settlements in a thin layer of an overconsolidated clay.

1.2.3 Scenario 1: Silos A and B are built simultaneously

Figure 1.9 illustrates how settlements are calculated in Scenario 1. The adopted notation is such that, e.g., $\Delta\sigma_{A3}$ is the stress increment caused by foundation A under the point E_3; Δe_3 is the changing in void ratio under point E_3. It contributes to settlements under E_3.

From Equation (1.1) and Figure 1.7 it follows (due to symmetry) that

$$\Delta\sigma_{A1} = \Delta\sigma_{A2} = \Delta\sigma_{B3} = \Delta\sigma_{B4} = \Delta\sigma, \tag{1.4}$$

$$\Delta\sigma_{A4} = \Delta\sigma_{B1} < \Delta\sigma_{B2} = \Delta\sigma_{A3} < \Delta\sigma, \tag{1.5}$$

which is true for all three scenarios. In Scenario 1, from Figure 1.9 it follows that $\Delta e_2 = \Delta e_3 > \Delta e_4 = \Delta e_1$ and the resulting settlements, calculated for our example in Figure 1.6b by dividing the clay layer into three 10 m thick sublayers, are

$$\rho_2 = \rho_3 = 141.3 \text{ cm} > \rho_4 = \rho_1 = 108.2 \text{ cm}.$$

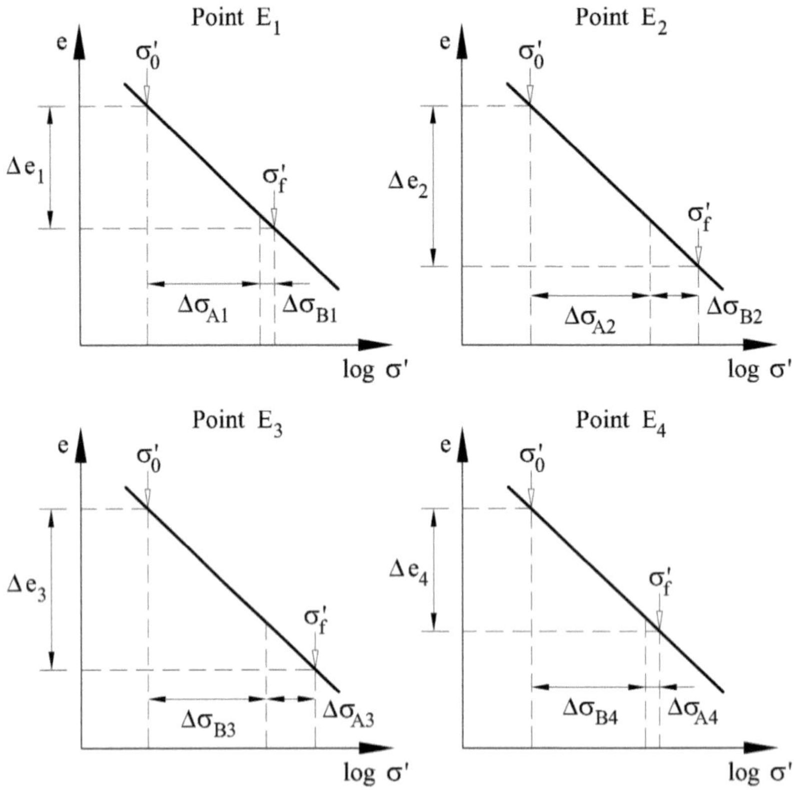

Figure 1.9 Schematic settlements in a thin layer in Scenario 1.

A numerical example of calculating the settlement under Point E_1 is presented in Table 1.1.

Table 1.1 Calculated settlements under Point E_1.

i (-)	z_i (m)	ΔH_i (m)	γ' (kN/m³)	σ'_0 (kN/m²)	$\Delta\sigma_{A1}$ (kN/m²)	$\Delta\sigma_{B1}$ (kN/m²)	$\Delta\sigma_{tot}$ (kN/m²)	$\Delta\rho$ (cm)
1	5	10	10	50	80	1	81	83.6
2	15	10	10	150	29	6	35	18.2
3	25	10	10	250	13	6	19	6.4
						Total settlement		108.2

where, in Table 1.1,

$$\Delta\sigma_{A1} = q \cdot 2 \cdot J(a = 10 \text{ m}, b = 5 \text{ m}),$$

$$\Delta\sigma_{B1} = q \cdot [2 \cdot J(a = 22 \text{ m}, b = 5 \text{ m}) - 2 \cdot J(a = 12 \text{ m}, b = 5 \text{ m})],$$

$$\Delta\sigma_{tot} = \Delta\sigma_{A1} + \Delta\sigma_{B1}.$$

1.2.4 Scenario 2: Silo B is built after Silo A

Figure 1.10 shows how settlements are calculated in Scenario 2. For each foundation point, σ'_A denotes the effective stress at the end of the Silo A construction; σ'_f is the final stress at the end of the Silo B construction.

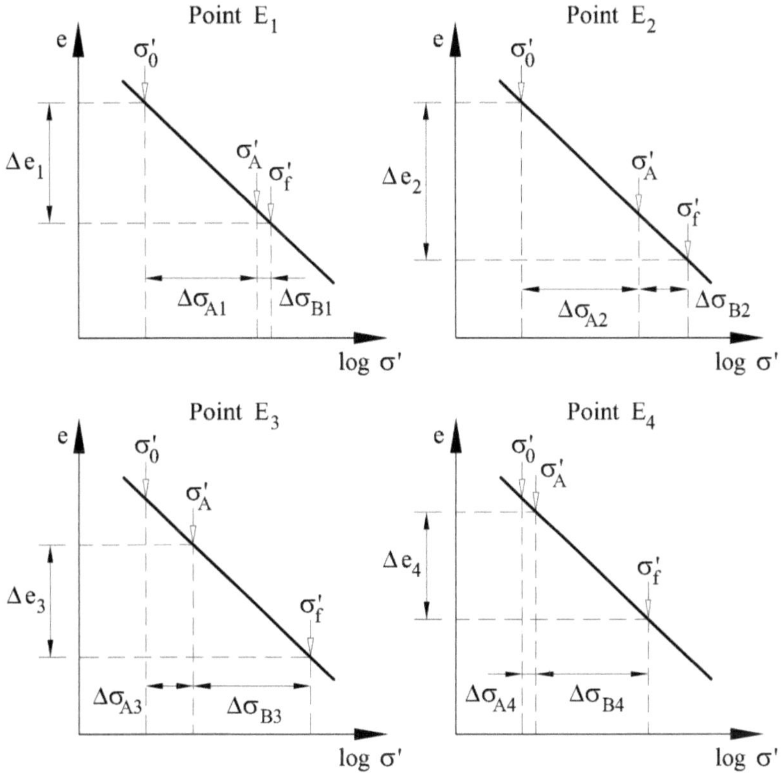

Figure 1.10 Schematic settlements in a thin layer in Scenario 2.

The stress increments in this scenario are still given by Equations (1.4) and (1.5), and the void ratio increments $\Delta e_2 > \Delta e_1$ are the same as in Scenario 1. The values of Δe_3 and Δe_4 are, however, different for the following reasons: (a) Silo B is built on levelled ground and its settlements do not include the ground deflections before its construction; (b) the same stress increment $\Delta\sigma_{B3} = \Delta\sigma_{B4} = \Delta\sigma$ causes a smaller increment Δe when applied at a larger initial stress (in this scenario σ'_A), due to a non-linear (logarithmic) nature of the stress-void ratio relationship (on the load axis, $\Delta\sigma_{B3}$ looks smaller than $\Delta\sigma_{B4}$). The resulting settlements for our case are

$$\rho_1 = 108.2 \text{ cm} < \rho_2 = 141.3 \text{ cm} \qquad \rho_3 = 72.8 \text{ cm} < \rho_4 = 101.3 \text{ cm}.$$

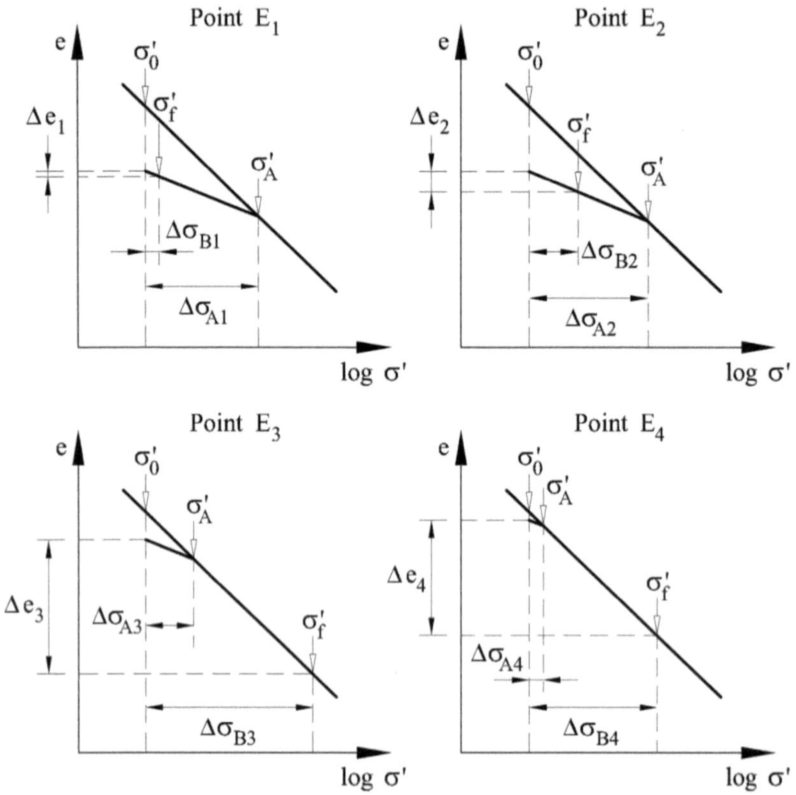

Figure 1.11 Schematic settlements in a thin layer in Scenario 3.

1.2.5 Scenario 3: Silo B is built after Silo A is removed

Figure 1.11 demonstrates how settlements are calculated in Scenario 3. In this scenario, the layer is first loaded from σ'_0 to σ'_A, then unloaded back to σ'_0, which turns the clay from normal to overconsolidated, with σ'_A being the preconsolidation stress. The larger this stress is, the smaller the increment Δe is going to be for the same stress increment $\Delta\sigma_{B3} = \Delta\sigma_{B4} = \Delta\sigma$.

The resulting settlements of Silo B calculated for our example in Figure 1.6b with three 10 m thick sublayers are

$$\rho_3 = 45.8 \text{ cm} < \rho_4 = 97.2 \text{ cm}.$$

1.2.6 Summary

The results of the above analysis are summarized in Figure 1.12. Symmetrical differential settlements Δ = 33 cm in Scenario 1 are explained by simple superposition of the stress bulbs under and outside the footings.

Scenario 2, however, produces asymmetric differential settlements which, at first glance, are not obvious. The asymmetry here has a simple explanation: Silo B is built on levelled ground and its settlements do not include the ground deflections which occurred before its construction. But in this case, why does this silo has differential settlements at all? Indeed, if the soil was linear elastic, the uniform contact pressure would result in zero differential settlements. The non-linear (logarithmic) stress-void ratio relationship, however, implies that the same stress increment causes a smaller settlement when applied at a larger initial stress (i.e. as we get closer to Silo A). In other words, the ground, which experiences a larger initial stress at the moment before loading, is denser and, therefore, less compressible.

The latter argument is valid for Scenario 3 as well with an important difference. In this case, at the moment before loading, the stress under Silo B is uniform. But the soil has a memory of larger stresses closer to where Silo A used to be. Therefore it is denser and less compressible at that side. Remarkably, this scenario produces the largest differential settlements (Fig. 1.12)!

Another important observation: increasing the distance between the silos by only 1 m would reduce the differential settlements by almost 30%!

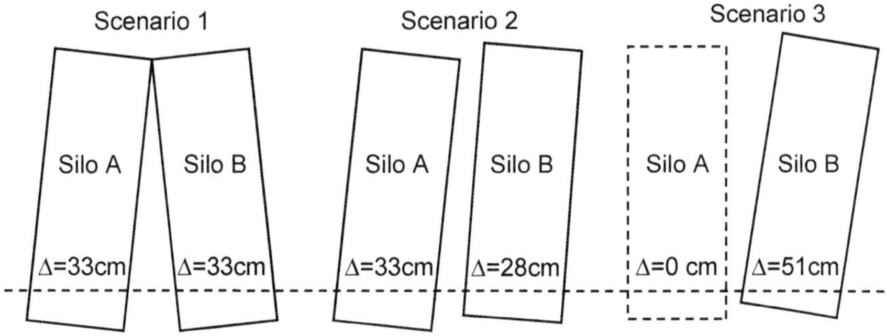

Figure 1.12 Summary of the silo interaction example.

1.3 The Analysis

The explored above silo model provides the understanding and the tools for a simplified analysis of the differential settlements of the Mexico City Metropolitan Cathedral.

1.3.1 Simplified model

A simplified geometry of the Cathedral and El Sagrario is presented in Figure 1.13a. The Aztec pyramid is assumed to have the same footprint as El Sagrario. The corresponding average contact pressures are shown in Figure 1.13b (the average height of the stone pyramid is assumed to be about 20 m, the unit weight of stone $\gamma = 25$ kN/m^3). The simplified soil profile is shown in Figure 1.14a. From a representative consolidation test in Figure 1.14b (curve of 1950):

$$e_0 = 7.0\,,\ \ C_c = \frac{e_2 - e_1}{\log(\sigma_2/\sigma_1)} = \frac{4.8-2.4}{\log(600/200)} = 5.0\,,\ \ C_e = \frac{3.1-2.6}{\log(100/10)} = 0.5\,.$$

The two curves in Figure 1.14b demonstrate how much decrease in the void ratio the clay experienced between 1950 and 1986 due to the consolidation caused by the exploitation of the aquifer (samples were taken from the same stratum with a 34 years interval). Note that the compression and swelling indexes were not affected by the consolidation.

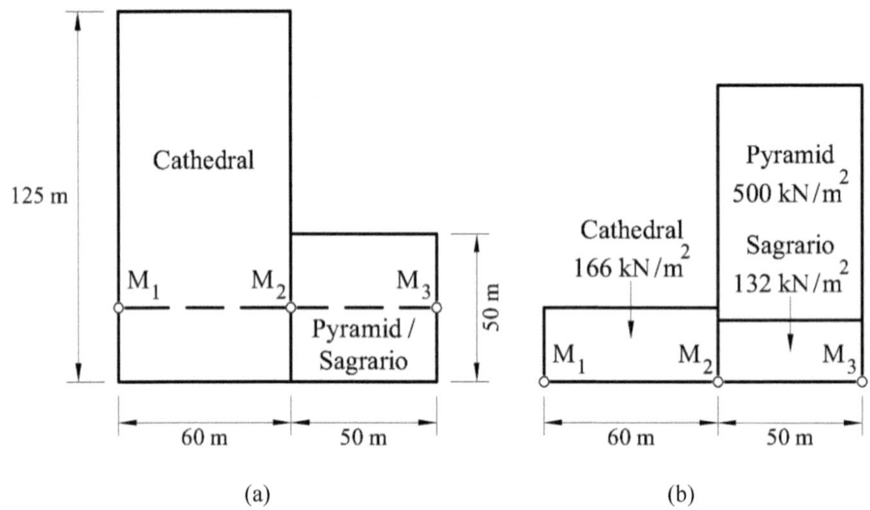

Figure 1.13 Simplified structural model: (a) plan; (b) contact pressure.

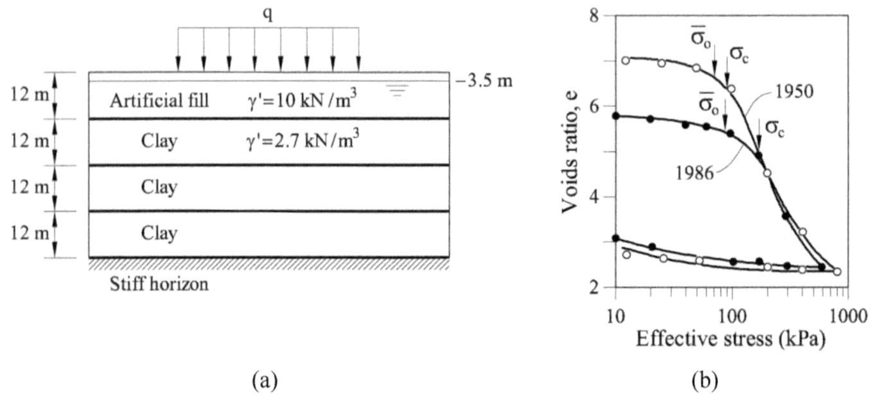

Figure 1.14 Geotechnical model: (a) simplified soil profile; (b) consolidation tests (after Méndez, 1991; Ovando-Shelley *et al.*, 2003).

1.3.2 Settlements due to consolidation

The final settlements due to consolidation in the clay layers are calculated at the points M_1, M_2 and M_3 (marked in Fig. 1.13a) by dividing the clay layer into three 12 m thick sublayers (Fig. 1.14a), and following the subsequent loading sequence (Fig. 1.13b):

- Stage I: the Pyramid ($\Delta\sigma_P$) is built and removed, then the ground is levelled and the Cathedral ($\Delta\sigma_C$) is built (Fig. 1.15);
- Stage II: the ground around the Cathedral is levelled and El Sagrario ($\Delta\sigma_S$) is built (Fig. 1.16).

In Figures 1.15 and 1.16, the adopted notation is again such that, e.g., $\Delta\sigma_{S3}$ is the stress increment caused by El Sagrario under point M_3; Δe_3 contributes to settlements under point M_3. For each point M_1 – M_3, σ'_0 and σ'_f denote the corresponding effective stresses at the beginning and at the end of the construction stage, respectively.

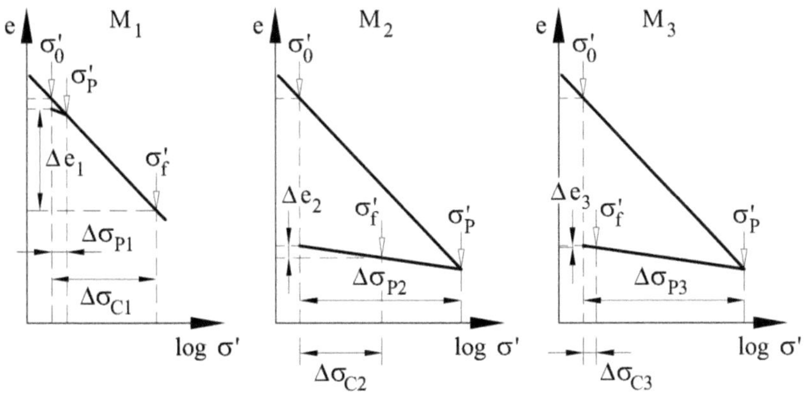

Figure 1.15 Calculation of settlements: Stage I.

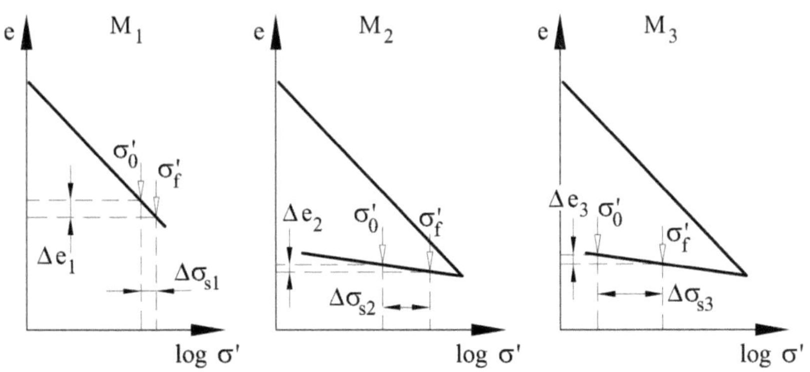

Figure 1.16 Calculation of settlements: Stage II.

The calculated settlements due to consolidation are given in Table 1.2. The calculated differential settlements of the Cathedral are overestimated, indicating that the loading history alone is not sufficient to explain them. The measured settlement is smaller, probably because a part of the settlement took place during the construction and was compensated in the process. The calculated differential settlements of El Sagrario are of the correct sign but are significantly underestimated, i.e. the loading history alone is not sufficient to explain them.

Table 1.2 Calculated settlements of the Cathedral and El Sagrario.

Settlement (cm)	Cathedral		El Sagrario	
	M_1	M_2	M_2	M_3
Stage I	279.7	29.5	-	-
Stage II	3.5	16.1	16.1	21.1
Total	283.2	36.7	16.1	21.1
Differential	246.5		5.0	
Measured in 1990	125.0		50.0	

1.3.3 Settlements due to a drop in the groundwater level

Let us now explore how a drop in the groundwater level (GWL) would affect the differential settlements of the Cathedral and El Sagrario by considering:

- Stage III: the global 4 m drop in the groundwater level (between 1972 and 1990).

Schematic calculation of settlements is illustrated in Figure 1.17. For each point $M_1 - M_3$, σ'_0 and σ'_f denote the corresponding effective stresses before and after the drop in the GWL, respectively. As is seen, this drop will increase the existing differential settlements, in spite of the fact that it causes a uniform stress increment $\Delta\sigma_G$ under both structures, because different points react to this stress increase differently.

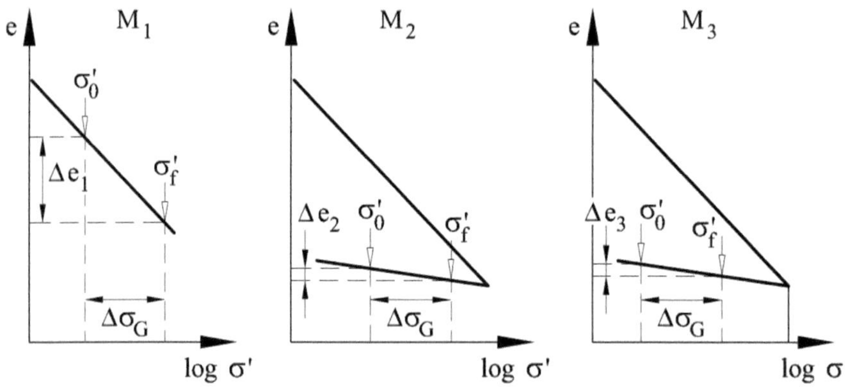

Figure 1.17 Calculation of settlements: Stage III.

At point M_1, the clay is normally consolidated, while at point M_2 it is overconsolidated, therefore M_2 settles less than M_1. Point M_2 settles less than M_3, in spite of the fact that the clay in both cases is overconsolidated. Below point M_2, however, the existing stress is much higher than under M_3. Therefore, the clay is more compacted and experiences smaller settlements for the same stress increment $\Delta\sigma_G$. Mathematically, this follows from the non-linear (logarithmic) nature of the stress-void ratio relationship.

The calculated settlements due to a drop in the groundwater level are given in Table 1.3. It appears that even if all the differential settlements of the Cathedral were compensated during its construction, the 4 m drop in the water level would be sufficient to cause the observed differential settlements. The calculated differential settlements of El Sagrario are again of the correct sign but are significantly underestimated, i.e. a global drop in the groundwater level is not sufficient to explain them.

Table 1.3 Calculated settlements due to a global 4 m drop in the groundwater level.

Settlement (cm)	Cathedral		El Sagrario	
	M_1	M_2	M_2	M_3
Stage III	133.5	11.5	11.5	14.3
Differential	122.0		2.8	
Measured in 1990	125.0		50.0	

One possible explanation for differential settlements of El Sagrario is the effect of the subway tunnel excavated along its eastern side (Fig. 1.4). The tunnel excavation creates a temporary drainage, causing a *local depression* in the groundwater level under point M_3. It can be shown that a 9 m deep depression leads to the observed 50 cm differential settlement of El Sagrario.

1.3.4 Discussion

The simplified geotechnical model of the problem of the Mexico City Metropolitan Cathedral cannot claim the exact prediction of the differential settlements. It focused only on the effects of the interaction of the neighbouring structures, loading history and the global and local drop in the groundwater level. The latter factor has been probably responsible for much higher than predicted differential settlements of El Sagrario. The smaller than predicted differential settlements of the Cathedral may be attributed to the correction measures carried out during its construction and later.

This simplified model, however, has been sufficient to achieve qualitatively correct predictions of the differential settlements of both the Cathedral and El Sagrario, of meaningful order of magnitude. It demonstrates the power of the simple geomechanical analysis in understanding the complex processes behind some geotechnical failures.

1.4 Mitigation Measures

Four proposals for correction of the differential settlements were considered in 1990. The first one – the addition of 1,500 piles – was rejected due to the lack of space for these piles and low chances for achieving sufficient bearing capacity. The second one – involving construction of 240 shafts some 60 m deep and 2.4 m in diameter – was rejected on economic grounds. The third one – construction of an impervious barrier along with water injection wells, in order to reestablish piezometric levels and reduce the effects of the regional subsidence – would require too much energy and water for continuous pumping.

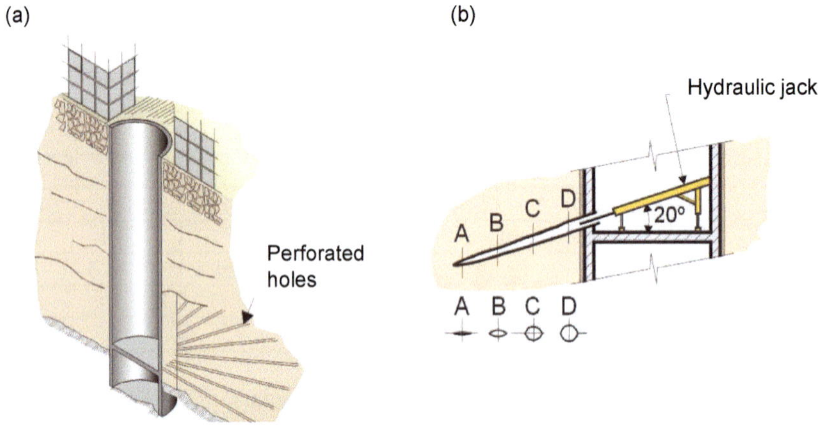

Figure 1.18 Underexcavation by soil extraction: (a) a vertical shaft; (b) perforation and collapse of a sub-horizontal borehole (after Santoyo and Ovando, 2002).

The fourth proposed method – underexcavation – was considered the most feasible. It consisted of removing soil by means of 10 cm diameter sub-horizontal borings using hydraulic jacks from a large diameter vertical shaft (Fig. 1.18).

Excavation was carried out below the foundation level, in plastic clays. Collapse of a perforated hole in a soil mass produces a surface settlement. As this hole collapses and a new one is penetrated, the settlement increases. Thus, the volume of the excavated soil (Fig. 1.19) gives rise to controlled settlement, allowing correction of distortion in the structure. By September 1999, the maximum achieved vertical correction was 88 cm.

The goal of the underexcavation has been to reestablish the conditions existing in the year 1934, to allow for comprehensive refurbishment of the Cathedral. The underexcavation procedures are likely to be repeated every 20 – 25 years. The structural deformation over these periods is expected to be small or negligible, thanks to the additional mitigation measure – grout injection into the upper clay layer, reducing its compressibility. About 5,190 m^3 of grout was injected, using hydraulic fracturing (Fig. 1.20), reducing the yearly displacements by an order of magnitude.

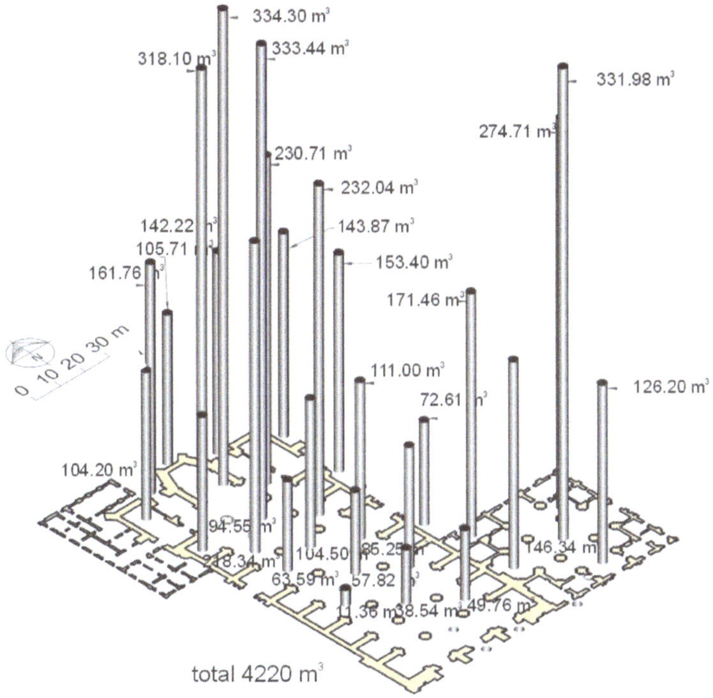

Figure 1.19 Location of the vertical shafts and the volume of the extracted soil (after Santoyo and Ovando, 2002).

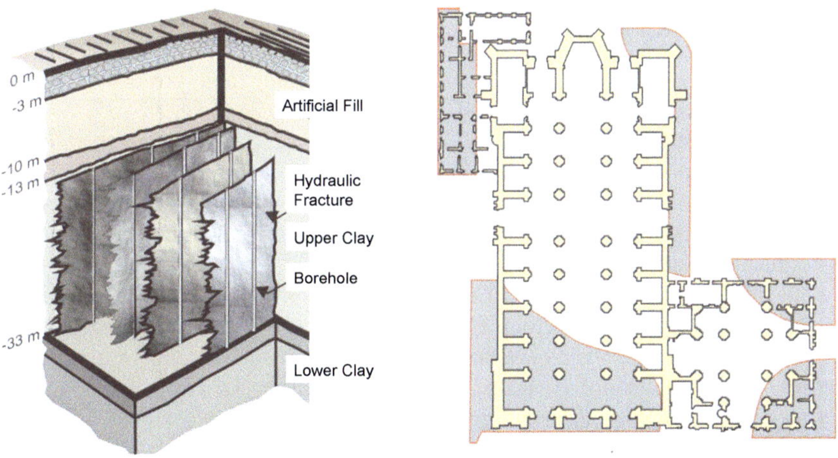

(a) (b)

Figure 1.20 Grout injection: (a) hydraulic fracturing; (b) injected areas (after Santoyo and Ovando, 2002).

1.5 Lessons Learned

1.5.1 Loading history

The history of the previous construction and the sequence of the current one will largely determine the pattern of the differential settlements of neighbouring structures by changing the compressibility of the soil foundation.

1.5.2 Distance between the neighbouring structures

Distance between the neighbouring structures is of crucial importance for the magnitude of the interaction effects. Due to the fast attenuation of stresses with distance, sometimes, a moderate increase in distance between the neighbouring structures can reduce the interaction effects to being small or negligible.

1.5.3 Regional subsidence

Though conventionally considered to be a source of the relatively uniform global settlements only, regional subsidence can cause differential settlements due to spatial variability in the compressibility of soil, resulting, e.g., from a specific loading history.

1.5.4 Do not mess with other people's gods!

Some people believe that the problems of the Cathedral are revenge of Aztec gods whose temple was destroyed to make way for the Cathedral. If this is the case, the gods used geotechnical engineering as their weapon!

References

Bozozuk, M. (1976) Tower silo foundations. *Canadian Building Digests, CBD-177*, Institute for Research in Construction, National Research Council Canada, 4 pp.

Guerra, S.Z. (1992) Severe soil deformations, leveling and protection at the Metropolitan Cathedral in Mexico City. *APT Bulletin* 24 (1/2), 28 – 35.

Lang, H.J., Huder, J., Amann, P. and Puzrin, A.M. (2007) *Bodenmechanik und Grundbau*. Springer Verlag, Berlin, 354 pp.

Méndez, E. (1991) Evolución de las propiedades de la arcilla de la Ciudad de México. *B. Sc. Thesis*, Escuela Superior de Ingeniería y Arquitectura, Instituto Politécnico Nacional, Mexico.

Ovando-Shelley, E., Romo, M.P., Contreras, N. and Giralt A. (2003) Effects on soil properties of future settlements in downtown Mexico City due to ground water extraction. *Geofísica Internacional* 42 (2), 185 – 204.

Santoyo, E. and Ovando, E. (2002) Paralelismo entre: la Torre de Pisa y la Catedral de México. *Proceedings of the International Workshop ISSMGE-Technical Committee TC36: Foundation Engineering in Difficult Soft Soil Conditions*. Mexico City, 34 pp.

Chapter 2

Unexpected Excessive Settlements: Kansai International Airport, Japan

TABLE OF CONTENTS

A.M. Puzrin et al., *Geomechanics of Failures*, DOI 10.1007/978-90-481-3531-8_2,

Chapter 2

Unexpected Excessive Settlements: Kansai International Airport, Japan

2.1 Case Description

2.1.1 Introduction

Kansai International Airport (KIA) in Osaka Bay, Japan (Fig. 2.1a) was singled out by the American Society of Civil Engineers as one of the "Monuments of the Millennium" – a designation awarded to the ten civil engineering projects deemed to have had the greatest positive impact on life in the 20th century. This is an even more remarkable achievement, considering that the first phase of the airport construction experienced some problems which, in a broader sense, could be characterized as a geotechnical failure. Built on a 1.25 km × 4 km man-made island (Fig. 2.1b), 5 km offshore at an average water depth of 18 m, this first phase experienced unexpected excessive settlements. These settlements were not correctly predicted either before or during the construction and caused delays, considerably increasing the cost of the 14 billion dollar project.

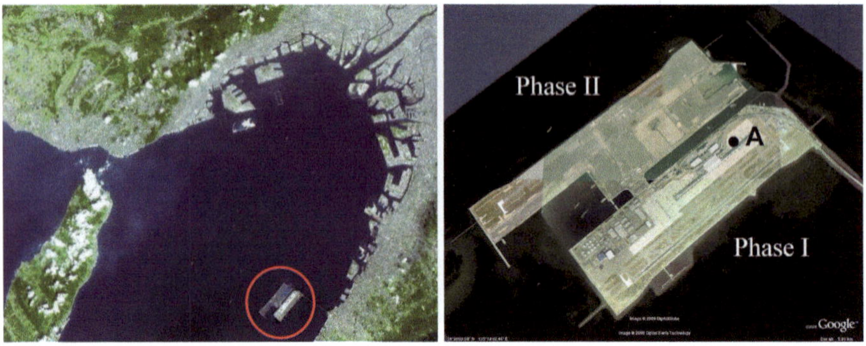

(a) (b)

Figure 2.1 Kansai International Airport (Google Earth ©): (a) Osaka Bay; (b) the islands of Phase I and Phase II.

2.1.2 Construction

Construction of the first phase island of Kansai International Airport started in January 1987 and finished in December 1991. Construction of the airport facilities followed and the airport began operations in September 1994. In the five years of the island's construction, more than 180 million m^3 of granular fill with a height of about 33 m were placed on the seabed, which consists of more than 1,200 m of sediments (Fig. 2.2). Only the upper 160 m, however, are considered to be compressible: the top 20 m are soft alluvial Holocene clays (Ma13), followed by the alternation of sand and clay layers of Pleistocene origin (Ma 7 – 12, "Ma" stands for marine clay).

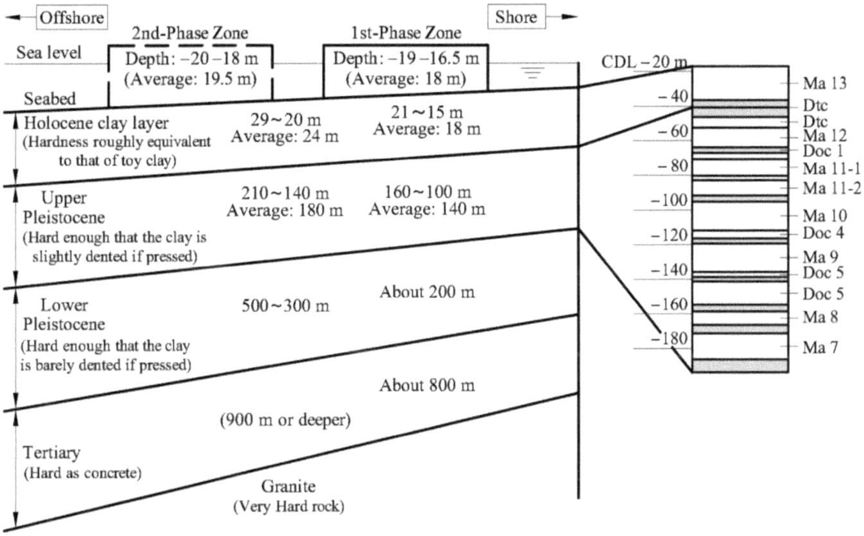

Figure 2.2 The soil profile of the seabed (after Akai *et al.*, 1995; Akai and Tanaka, 1999: © 1999 Taylor and Francis Group. Used with permission; KALD, 2009): dark layers-sand, white-marine clays Ma7 – Ma13.

The construction process included the following stages. First, the top 20 m of the seabed (Holocene clays) were treated within the design island area by installing vertical sand drains to accelerate compaction under the backfill. Next, the perimeter seawalls were built (Fig. 2.3). Subsequently, the land reclamation took place, in which the granular fill, taken from a number of excavations in the Osaka area (they practically levelled a couple of large hills), was placed within the seawalls up to a depth of about 3 m below the water level using bottom-dump barges. The final step was accomplished by means of four large barges, anchored inside the seawalls, which transferred the fill brought by the smaller barges from across the bay, to bring the island to the required 4 m above the water level. This height is to guarantee that the airport will not be swamped by high tides brought by typhoons that hit the coast of Japan every September.

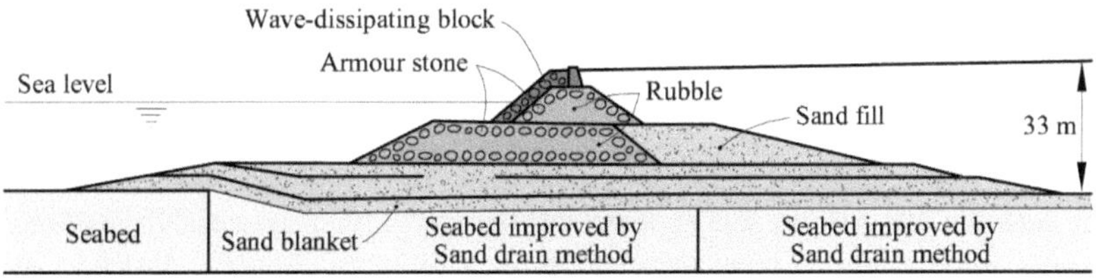

Figure 2.3 The cross-section of a seawall (after KALD, 2009).

Two important notes on the construction process: (1) no measures have been taken to accelerate the consolidation settlements in the lower (Pleistocene) clay strata, due to its considerable depth (instead, a special effort was made to predict the settlements as accurate as possible); (2) once the final step was complete, there was no way to add the fill on top of the island using the same method. The problem was that the barges could not enter the island's interior anymore. Combination of these two factors shows that the planning and design processes did not allow for the risk of extra settlement due to consolidation of the Pleistocene clay strata. No construction process was devised which could cope with these settlements.

2.1.3 The history of settlements

Thanks to the vertical drains, the top 20 m of the Holocene clay reached almost 90% of its final 6 m settlement during the construction (Handy, 2002). These settlements were accounted for in the design, being compensated by an additional 6 m thick layer of fill and additional height of the seawalls. What was apparently not fully accounted for were the excessive settlements of the Pleistocene clays and their slow accumulation in time (Fig. 2.4). By 1999, immediate settlements of about 1 m magnitude were recorded followed by additional 5 m of settlement and continue to increase at a rate of about 15 cm per year. Adding extra layers of fill during the construction compensated only for immediate settlements and a small part of the consolidation settlement of the Pleistocene sediments.

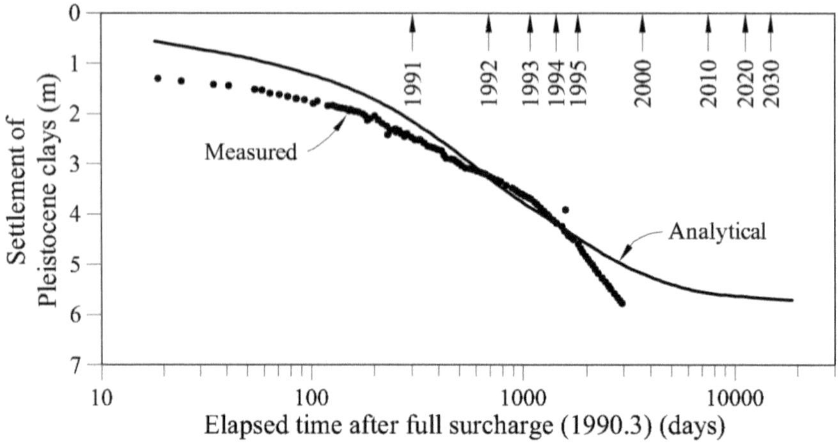

Figure 2.4 Settlements of the island (at point A in Fig. 2.1b) due to consolidation of Pleistocene clays (after Endo *et al.*, 1991; Akai and Tanaka, 1999: © 1999 Taylor and Francis Group. Used with permission).

The original design did not anticipate these developments. As soon as it was clear that the Pleistocene clays were the source of these excessive settlements, attempts were made to re-evaluate them as shown in Figure 2.4 (Endo *et al.*, 1991). These attempts were based on field measurements at the beginning of the construction and failed to provide a reasonably accurate prediction. They were continuously corrected, but each time a prediction was made, the island provided a new "surprise". First, it was a sizeable immediate settlement. Next, it was a much slower than expected rate of settlement at the beginning of the consolidation. Finally, these settlements did not seem to slow down as much as expected towards the end of the consolidation.

2.1.4 The problem

Many different factors affected the settlement prediction of the Kansai International Airport. This chapter will focus on just three of them, mentioned above: the immediate settlement, the slower settlement rate in the beginning, and the faster rate towards the end of consolidation of the upper Pleistocene clays.

Immediate settlements are caused by the three-dimensionality of the real problem, as compared to the one dimensional consolidation theory, in particular, by non-negligible lateral strains. They are also affected by the compressibility of sand layers where dissipation of pore water pressures takes place very quickly.

The rate of consolidation is determined to a large extent by the length of the drainage path. Conventionally, a sand layer between two clay layers is considered to work as a drain. In the case of the Pleistocene strata, however, some of the sand layers proved to be lenses, entirely enclosed within the clay layer, which was confirmed by very slow dissipation of pore water pressures measured in the sand layers 10 years after the land reclamation (Fig. 2.5).

Finally, the consolidation process is not the only one controlling the rate of settlements in clay. Even when the excess pore water pressure has completely dissipated, the settlement continues, which is called creep or *secondary compression*. Towards the end of primary consolidation, a contribution of the secondary compression becomes more pronounced and may produce significant increases in settlements long after the primary consolidation is over.

All these factors are rather difficult to quantify accurately in advance of the construction. This is not, however, good news for designers. Is there any way to resolve this dilemma?

2.1.5 The observational method

The major problem with land reclamation projects of this scale is that it is almost impossible to provide an accurate prediction of the rates of settlement based solely on the results of site investigation and laboratory consolidation tests. There are two major reasons for that: (1) large spatial variability of soil properties and drainage geometry and (2) laboratory tests often produce the values of the coefficient of consolidation c_v and secondary compression C_α within two orders of magnitude from the field values. Therefore, such estimates can only be used as initial conditions for design.

In such a case, the design should be left flexible to accommodate changes, as construction proceeds. These changes are based on the continuous monitoring of significant field parameters and on inverse analysis of the field measurements. This back calculation allows for the model parameters to be updated using real field data and then utilized for the next stages of analysis and design. In geotechnical engineering this approach is called *the observational method.* The purpose of this chapter is to demonstrate a simplified back-calculation procedure for the field data from the KIA.

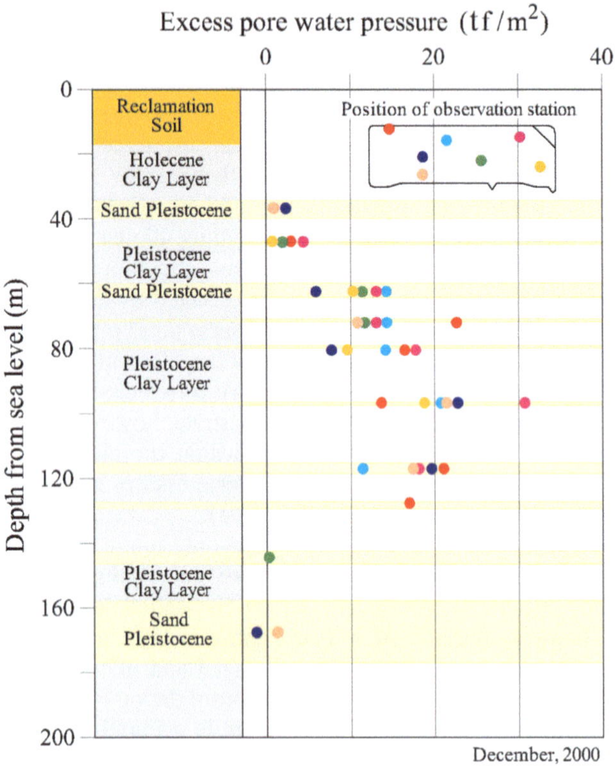

Figure 2.5 Drainage in the sand layers: 1 tf/m^2 = 9.8 kPa (after KALD, 2009).

2.2 *The One-Dimensional Theory*

Marine deposits near the coast and big river deltas are often layered systems due to their alluvial nature. In general, they are built of alternating sand and clay layers, whose thickness depends on the geological history. The settlements of the saturated sand and clay materials are governed by different phenomena. Settlements of saturated sand layers are normally immediate settlements, provided that they work as open drains. Settlement of saturated clay layers is a time-dependent phenomenon which is governed by the following processes:

- primary consolidation – decrease in the volume of pores due to the flow of water caused by the dissipating pore water pressure gradient;
- secondary compression – decrease in the pore volume due to creep.

The total settlement of these layered systems in a one-dimensional problem is built, therefore, out of these three distinct components:

$$\rho_T(t) = \rho_I + \rho_C(t) + \rho_S(t), \tag{2.1}$$

where ρ_I is the immediate settlement (sand);
$\rho_C(t)$ is the consolidation settlement (clay);
$\rho_S(t)$ is the creep settlement (clay).

2.2.1 Immediate settlement

Immediate settlements in saturated clay layers in a one-dimensional problem should be zero, due to the small pores and incompressibility of the pore water which cannot leave the pores immediately. In reality, however, there are always some immediate settlements caused by the three-dimensionality of the real problem, which involves non-negligible lateral strains. Because of these lateral strains, some vertical settlements also occur without any change in the total volume, as required by the incompressibility condition.

In the layered sand-clay systems, however, some immediate settlements occur, even in a one-dimensional problem. This is due to the compressibility of the sand layers, where pores are sufficiently large to allow for almost immediate dissipation of pore water pressures. These immediate settlements due to a stress increase of $\Delta\sigma$ in a thin sand sublayer of thickness *H*, can be calculated as

$$\rho_I = H\frac{\Delta\sigma}{M_E^{\text{sand}}} = H\frac{(1+\nu)(1-2\nu)}{E(1-\nu)}\Delta\sigma, \tag{2.2}$$

where M_E^{sand} is the one-dimensional compression modulus of sand;
E is the Young modulus of sand;
$\nu = 0.2 \div 0.3$ is the Poisson ratio of sand.

Note that for clay, this formula would produce a zero settlement due to the incompressibility condition $\nu = 0.5$.

Immediate settlements take place during construction. Therefore, though important at the design stage of land reclamation (for correct estimates of the required fill volume), they do not affect the long-term behaviour of the structure and have to be properly excluded from the analysis.

2.2.2 Settlement due to one-dimensional consolidation

Calculation of the final settlement due to consolidation has been discussed in Section 1.2 (Chapter 1). By the end of the consolidation, a total stress increase of $\Delta\sigma$ in a thin sublayer of thickness *H* will produce an equal effective stress increment $\Delta\sigma' = \Delta\sigma$. For normally consolidated clays, this will result in a final settlement of

$$\rho_{\text{inf}} = \frac{H}{M_E^{clay}}\Delta\sigma \quad \text{or} \quad \rho_{\text{inf}} = \frac{H}{1+e_0}C_c \log\frac{\sigma_0' + \Delta\sigma}{\sigma_0'}, \tag{2.3}$$

where σ_0' is the effective normal vertical stress;

e_0 is the in-situ void ratio before the construction;

C_c is the compression index;

M_E^{clay} is the one-dimensional compression modulus of clay.

C_c and M_E^{clay} are related as follows:

$$\frac{1}{M_E^{\text{clay}}} = \frac{C_c}{1+e_0} \frac{\log(\sigma_0' + \Delta\sigma') - \log(\sigma_0')}{\Delta\sigma'}. \tag{2.4}$$

The development of the consolidation settlement in time is schematically presented in Figure 2.6a. The total stress increment $\Delta\sigma$ (constant in time and uniformly distributed with depth) is first entirely taken by the pore water. This causes a pressure gradient between the clay layer and draining boundaries, producing a pore water flow towards these boundaries. In the process, the excess pore pressure $\Delta u(t)$ dissipates and more load is transferred to the soil skeleton via the effective stresses $\Delta\sigma'(t) = \Delta\sigma - \Delta u(t)$, causing its compression and settlements $\rho_C(t)$.

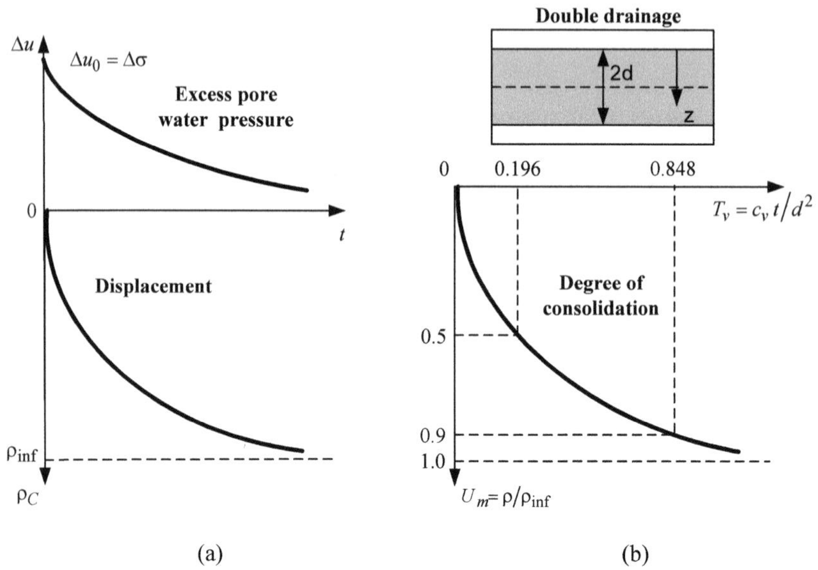

Figure 2.6 One-dimensional consolidation: (a) the process; (b) analytical solution.

The analytical solution to this problem, derived by Terzaghi (1943), is presented in a non-dimensional form in Figure 2.6b:

$$U_m(T_v) = \frac{\rho_C}{\rho_{\text{inf}}} = 1 - \sum_{m=0}^{\infty} \frac{2}{M^2} \exp(-M^2 T_v), \tag{2.5}$$

where $U_m = \frac{\rho_C}{\rho_{inf}}$ is the average degree of consolidation;

$T_v = \frac{c_v \cdot t}{d^2}$ is the non-dimensional time factor;

d is the drainage path; in a layer with double drainage, half of its thickness H;

$c_v = \frac{k \cdot M_E^{clay}}{\gamma_w}$ is the coefficient of consolidation;

k is the permeability coefficient;

γ_w is the unit weight of the water;

$M = \frac{\pi}{2}(2m+1)$, where $m = 0, 1, 2, ..., \infty$.

Relationship (2.5) can be quite accurately approximated by the following two analytical functions (Terzaghi, 1943):

$$T_v \approx \frac{\pi}{4} U_m^2, \quad \text{for } U_m \le 0.526; \tag{2.6}$$

$$T_v \approx -0.933 \cdot \log(1 - U_m) - 0.085, \quad \text{for } U_m > 0.526. \tag{2.7}$$

Using these relationships, we can describe the variation of consolidation settlement in time analytically:

$$\rho_C(t) = \rho_{inf} \cdot U_m = \rho_{inf} \cdot \sqrt{\frac{4 \cdot T_v}{\pi}} = \rho_{inf} \cdot \frac{2}{d} \sqrt{\frac{c_v \cdot t}{\pi}} \quad \text{for } U_m \le 0.526; \tag{2.8}$$

$$\rho_C(t) = \rho_{inf} \cdot U_m = \rho_{inf} \cdot \left(1 - 10^{-\frac{c_v t / d^2 + 0.085}{0.933}} \right) \quad \text{for } U_m > 0.526. \tag{2.9}$$

For layered strata, the total settlement is calculated as the sum of the settlements of individual clay layers:

$$\rho_C(t) = \sum_{i=1}^{n} \rho_C^i(t). \tag{2.10}$$

2.2.3 Secondary compression (creep) settlements

Creep settlements begin together with the primary consolidation settlements, but become dominant only towards the end of the primary consolidation (Fig. 2.7a) and can be predicted using the formula (e.g., Mesri and Vardhanabhuti, 2005):

$$\rho_S(t) = \frac{C_\alpha}{1+e_0} H \log\left(\frac{t}{t_p}\right), \tag{2.11}$$

where C_α is the coefficient of secondary compression;

t_p is the assumed beginning of the secondary compression, defined in Figure 2.7a at the point, where the experimental curve starts deviating from the theoretical primary consolidation line.

2.2.4 Total settlements

Using Equations (2.8) – (2.10), the variation of total settlement in time can be predicted using the following formulae:

$$\rho_T(t) = \rho_I + \rho_{\text{inf}} \frac{2}{d}\sqrt{\frac{c_v \cdot t}{\pi}} \quad \text{for } 0 \le t \le 0.217\frac{d^2}{c_v}; \tag{2.12}$$

$$\rho_T(t) = \rho_I + \rho_{\text{inf}}\left(1 - 10^{-1.072 c_v t/d^2 - 0.091}\right) \quad \text{for } 0.217\frac{d^2}{c_v} < t \le t_p; \tag{2.13}$$

$$\rho_T(t) = \rho_I + \rho_{\text{inf}}\left(1 - 10^{-1.072 c_v t/d^2 - 0.091}\right) + \frac{C_\alpha}{1+e_0} H \log\left(t/t_p\right) \quad \text{for } t > t_p, \tag{2.14}$$

where ρ_I is the initial settlement;

t_p is the assumed beginning time of the secondary compression.

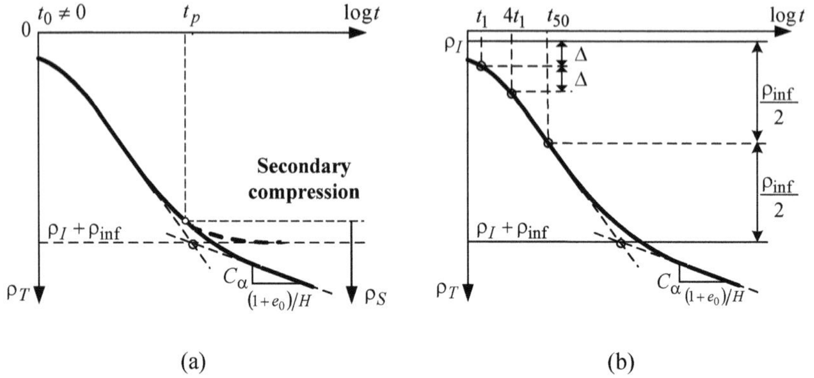

Figure 2.7 Semi-logarithmic settlement-time plot: (a) secondary compression; (b) graphic procedure for back-calculation of model parameters.

2.2.5 Inverse analysis of the settlement data

The parameters in Equations (2.12) – (2.14) can be easily back-calculated from the measured settlement data using the graphic procedure in Figure 2.7b:

- Measure the settlement Δ between any two early time moments t_1 and $4t_1$, and add it to $\rho(t_1)$. Using the fact that the initial part of the settlement curve is parabolic (Eq. 2.8), this gives the initial settlement

$\rho_I = \rho(t_1) + [\rho(t_1) - \rho(4t_1)] = 2\rho(t_1) - \rho(4t_1)$;

- Draw two tangential lines to straight portions of the primary consolidation and secondary compression curves. Their intersection defines $\rho_I + \rho_{\inf}$. Using ρ_I found above, this gives $\rho_{\inf}$;
- the slope of the tangent to the secondary compression curve is C_α;
- the point on the curve with $\rho(t_{50}) = \rho_I + \rho_{\inf}/2$, at which the first half of the final consolidation settlement was reached ($U_m = 0.5$), gives t_{50}, so that $c_v = 0.196\, d^2 / t_{50}$ (see Fig. 2.6b).

2.3 The Analysis

The one-dimensional theory presented above provides tools for a simplified analysis of the unexpected excessive settlements of Kansai International Airport.

2.3.1 Simplified model

The upper clay layer in Figure 2.8a (Holocene Clay MA13) consolidated very quickly thanks to the vertical drains and its settlement is assumed to be immediate. The nine upper Pleistocene clay layers, sandwiched between the ten sand layers (KALD, 2009) are assumed, for simplicity, to have the same thickness H = 12.0 m (Fig. 2.8b).

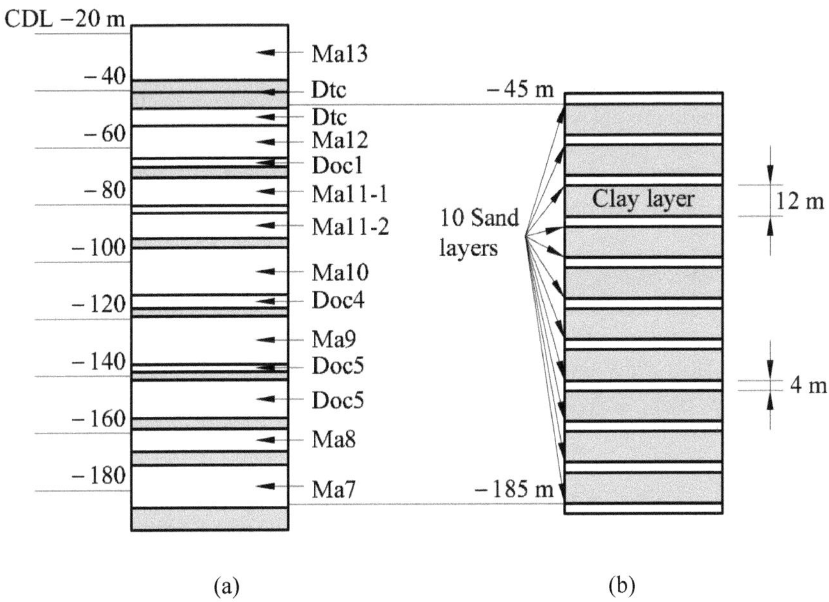

Figure 2.8 Upper Pleistocene soil profile: (a) real (after Akai and Tanaka, 1999: © 1999 Taylor and Francis Group. Used with permission); (b) simplified.

As mentioned in Section 2.1.2, as soon as it became clear that the Pleistocene clays are the source of excessive settlements, attempts were undertaken to make a prediction of their future developments. One of these predictions, by Endo *et al.* (1991), provided an analytical solution, shown in Figure 2.4 by a solid line. Our simplified model will reproduce this solution if the following assumptions are made with respect to the geometry and material properties:

- the height of the 4.0 km$\times$1.25 km island above the seabed is $h = 33$ m, out of which $h_w = 29$ m are below the sea level;
- the unit weights of soil are $\gamma_{\text{island}} = 21$ kN/m^3, $\gamma_{\text{seabed}} = 18$ kN/m^3;
- the overconsolidation ratio OCR of Pleistocene deposits grows linearly with depth (Akai *et al.*, 1995), therefore the lower Pleistocene clay remains overconsolidated in the process of land reclamation and can be considered incompressible;
- the Pleistocene clay is considered to be normally consolidated with consolidation parameters $e_0 = 1.5$, $C_c = 0.6$, $c_v = 1.67\times10^{-7}$ m^2/s roughly corresponding to the results of consolidation tests (Fig. 2.9);
- the upper Pleistocene clay of total thickness $H = 108$ m is built of nine 12 m thick sublayers interlaid by eight 4 m thick sand drains so that the average length of the vertical drainage path $d = 6.0$ m;
- sand layers are also present at the lower and upper boundaries of the 108 m thick Pleistocene clay layer;
- neither an initial settlement nor a settlement due to the secondary compression were considered.

2.3.2 The original prediction

In this section, using our simplified model, we reproduce the original prediction of Endo *et al.* (1991) represented by the solid curve in Figure 2.4.

The Ma13 Holocene Clay layer consolidated before the end of construction. This immediate settlement was approximately 6 m, and its increase in time is insignificant.

The final settlement due to consolidation of the Pleistocene clay layers can be calculated using formula (2.3) for each of the nine clay layers and summing their settlements:

$$\begin{aligned}\rho_{\text{inf}} &= \sum_{i=1}^{i=9} \Delta\rho^i_{\text{inf}} = \sum_{i=1}^{i=9} \Delta H \frac{C_c}{1+e_0} \log\left(\frac{\sigma'_{0i} + \Delta\sigma}{\sigma'_{0i}}\right) = \\ &= \sum_{i=1}^{i=9} 12 \frac{0.6}{1+1.5} \log\left(\frac{8D_i + 403}{8D_i}\right) = 5.6 \text{ m},\end{aligned} \tag{2.15}$$

where $\Delta\sigma = \gamma_{\text{island}}(h - h_w) + (\gamma_{\text{island}} - \gamma_w)h_w = 21\times4 + 11\times29 = 403$ kPa;

$\sigma'_0 = \gamma'_{\text{seabed}} D_i = 8D_i$ is the geological stress;

$D_i = 31\text{ m} + 16\text{ m}\cdot(i-1)$ is the depth of the center of the i-th clay layer from the seabed surface.

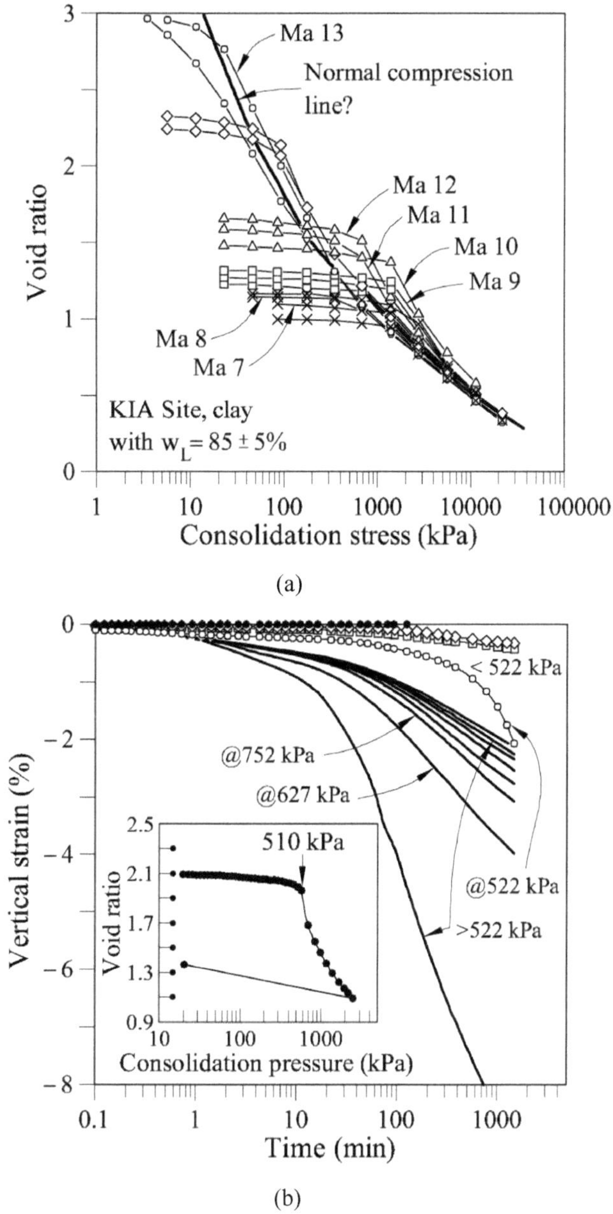

Figure 2.9 Consolidation test results for Pleistocene clays (after Akai and Tanaka, 1999: © 1999 Taylor and Francis Group. Used with permission): (a) consolidation curves; (b) settlement in time curves for M12. Here @522 kPa denotes the curve at the load close to the preconsolidation pressure of 510 kPa. Its rate of settlement is significantly higher than that of both, the overconsolidated (load < 522 kPa) and the normally consolidated (load > 522 kPa) curves, indicating destructuring, typical for aged clays.

This final 5.6 m consolidation settlement of the Pleistocene clay was also predicted by Endo *et al.* (1991) and is most likely to be a correct estimate (Akai and Tanaka, 2005). The problem of this prediction is different – it is not the value of the final consolidation settlement but the development of this settlement in time and what happens after the consolidation is over. Indeed, according to Endo *et al.* (1991), half of the Pleistocene clay settlement in all clay layers should have occurred after 490 days and 90% of it after 2,120 days, which can be also reproduced by our simplified model:

$$t_{50} = \frac{T_{v(50)} d^2}{c_v} = \frac{0.196 \times (6.0)^2}{1.67 \times 10^{-7}} \left(\frac{1}{24 \times 3600} \right) = 490 \text{ days},$$

$$t_{90} = \frac{T_{v(90)} d^2}{c_v} = \frac{0.848 \times (6.0)^2}{1.67 \times 10^{-7}} \left(\frac{1}{24 \times 3600} \right) = 2{,}120 \text{ days}.$$

In reality, however, the consolidation continued much longer. In 1999 (almost 3,200 days after the end of construction), the settlement already exceeded 6 m (the total settlement exceeded 12 m) and kept increasing at an average rate of about 15 cm per year (Fig. 2.4).

2.3.3 Correction for the initial settlement

Starting from the early stages of consolidation, it became clear that settlement of the Pleistocene deposits was much higher than expected (Fig. 2.10a). One possible reason for that could be a higher rate of consolidation, but the trend in Figure 2.10a is opposite – the rate of settlement is slower than predicted.

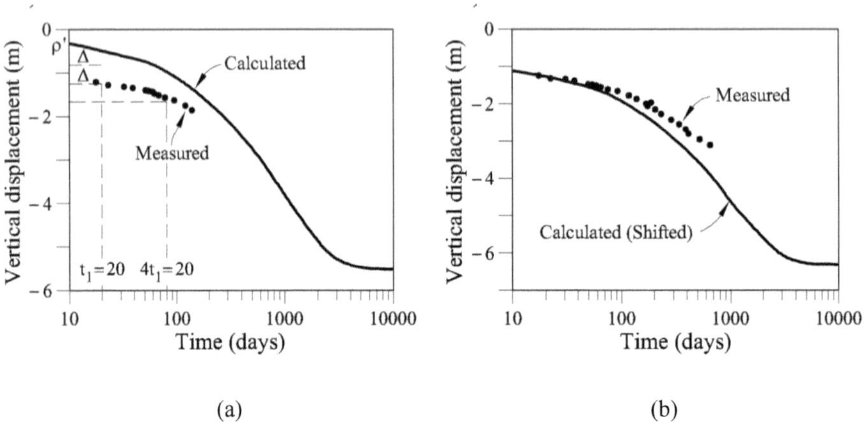

Figure 2.10 Correction for the initial settlement: (a) derivation of parameters; (b) corrected prediction.

Therefore, the most likely reason for higher settlements is an initial settlement. It can be estimated using the procedure in Figure 2.6b:

- at $t_1 = 20$ days: $\rho_{t1} = 1.25$ m;
- at $t_2 = 4t_1 = 80$ days: $\rho_{t2} = 1.60$ m;
- therefore, $\rho_I = 2\rho(t_1) - \rho(4t_1) = 2\times 1.25 - 1.60 = 0.90$ m.

The corrected prediction (Eq. 2.12), accounting for this initial settlement $\rho_I = 0.90$ m, is shown in Figure 2.10b. While giving a good fit to the measured data up to the first 60 days, it predicts a much higher rate of consolidation later on.

2.3.4 Correction for the length of the drainage path

The rate of consolidation is governed by the consolidation coefficient c_v and the average length of the drainage path d (or, in our model, the number n of draining sand layers). Because dissipation of the excess pore water pressure in some of the sand layers was very slow (Fig. 2.5), it is most probable that the average length of the drainage path d was larger than assumed from the geometry. Also, because d in the formula for the time factor T_v is squared and c_v is not, an inaccuracy in d affects the rate of consolidation stronger.

We assume that the final settlement due to consolidation of Pleistocene clay $\rho_{inf} = 5.6$ m was correctly predicted in Equation (2.15) (see also Akai and Tanaka, 2005). Then the measured time t_{50} of the 50% of consolidation settlement (i.e. at the total settlement of $\rho_{50} = \rho_I + \rho_{inf}/2 = 3.70$ m) was 1,000 days (Fig. 2.11a) and not 490 days, as predicted by Endo *et al.* (1991).

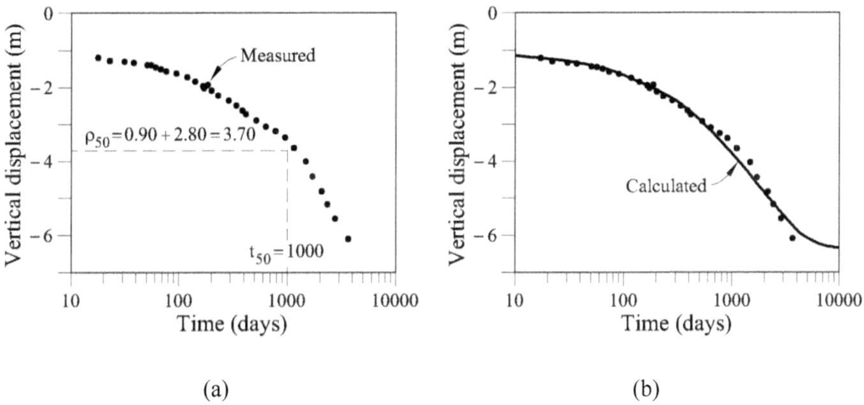

Figure 2.11 Correction for the length of the drainage path: (a) derivation of parameters; (b) corrected prediction.

The average length of the drainage path d and the number of equivalent clay layers n should be then adjusted as follows:

$$d = \sqrt{\frac{t_{50}\cdot c_v}{T_{v(50)}}} = \sqrt{\frac{1,000\times 24\times 3,600\times 1.67\times 10^{-7}}{0.196}} = 8.57 \text{ m} > 6 \text{ m}, \tag{2.16a}$$

$$n = \frac{H}{2d} = \frac{108}{2 \times 8.57} = 6.3 < 9. \tag{2.16b}$$

The corrected prediction (using Eqs. (2.12) and (2.13)), which accounts for this modified average length of the drainage path $d = 8.57$ m, is shown in Figure 2.11b. While giving a good fit to the measured data up to 2,000 days, later on it predicts a lower rate of settlement.

2.3.5 Correction for the secondary compression

The most likely reason for continuing excessive settlements at the final stages of consolidation is the secondary compression. In particular, Akai and Tanaka (1999) noted a very high rate of post-consolidation settlement when effective stress slightly exceeds the pre-consolidation pressure (Fig. 2.9b, the curve @522 kPa), and related this to the phenomenon of destructuring of aged clays. In order to be able to derive parameter C_α using the procedure in Figure 2.7b, we need more data points in time. However, if the latest prediction of the consolidation curve (corrected for initial settlement and the drainage path length) were reliable, then an alternative procedure could be applied (Fig. 2.12a).

According to this procedure, t_p is taken as the moment in time when the measurements start to deviate from the theoretical consolidation line, i.e. in our case: t_p = 1,800 days. Parameter C_α is then obtained from the difference $\Delta\rho_S$ between the predicted (with $C_\alpha = 0$) and measured displacements at some time $t > t_p$, e.g., for t = 3,200 days, $\Delta\rho_S = 0.37$ m, so that

$$C_\alpha = \frac{\Delta\rho_S\left(1+e_0\right)}{H\log\left(t/t_p\right)} = \frac{0.37\times\left(1+1.5\right)}{108\times\log\left(3,200/1,800\right)} = 0.034\,, \quad \frac{C_\alpha}{C_c} = \frac{0.034}{0.6} = 0.057.$$

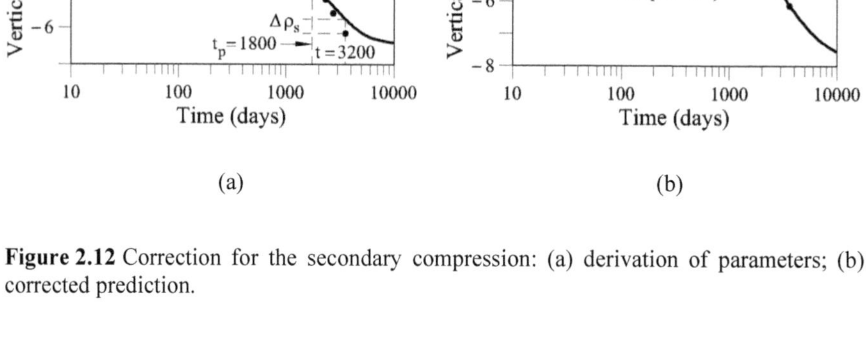

Figure 2.12 Correction for the secondary compression: (a) derivation of parameters; (b) corrected prediction.

The latter ratio of 0.057 is reasonable but somewhat high for clays, most probably due to the phenomenon of destructuring of aged clays, observed by Akai and Tanaka (1999) for the upper Pleistocene clays.

The corrected prediction (Eqs. (2.12) – (2.14)), accounting for the secondary compression with t_p = 1,800 days and $C_\alpha = 0.034$, is shown in Figure 2.12b. This prediction gives an excellent fit to the settlement data measured so far, considering the enormous simplifications which were introduced into the model.

2.3.6 Total predicted displacement

Assuming 50 years for the airport lifetime, we obtain the creep settlement:

$$\rho_S = \frac{C_\alpha}{1+e_0} H \log\left(\frac{t}{t_p}\right) = \frac{0.034}{1+1.5} \times 108 \times \log\left(\frac{50 \times 365}{1,800}\right) = 1.5 \text{ m}.$$

The total predicted displacement (after 50 years) for the Pleistocene layer then becomes:

$$\rho_T^{Pl} = \rho_I + \rho_{\text{inf}} + \rho_S = 0.9 + 5.6 + 1.5 = 8.0 \text{ m}.$$

Adding the settlement of the Holocene clay layer, we obtain (Fig. 2.13):

$$\rho_T = 8.0 + 6.0 = 14.0 \text{ m}.$$

The latest prediction of the 50 years settlement (Akai and Tanaka, 2005) is 14.3 m. If the design of the island was produced using the originally predicted 11.6 m settlement, the planned 4 m embankment over the sea level would, over the years, become reduced to just 1.6 m. This would not be sufficient to withstand the high tides brought by typhoons.

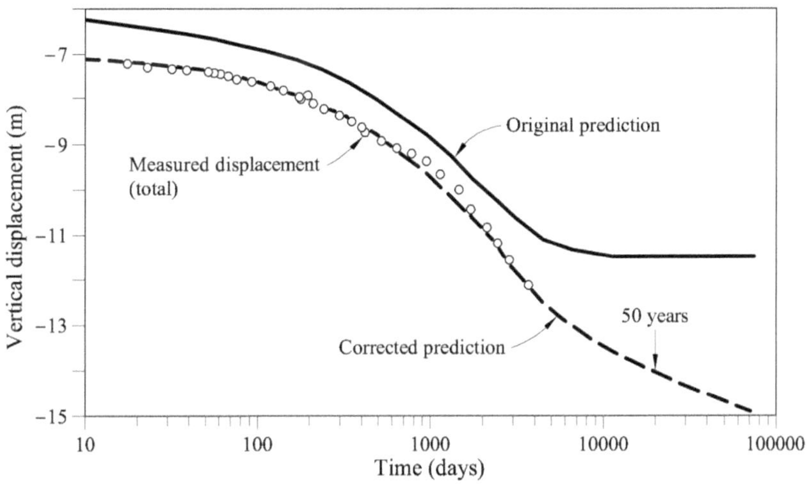

Figure 2.13 Comparison between the original and updated predictions of the total settlement.

2.3.7 Discussion

The simplified geotechnical model of the Kansai International Airport settlement is one-dimensional and therefore cannot account for the spatial variability of geometry, loading and soil properties, as well as for the three-dimensional effects of stress and strain distribution. It also focused only on the following phenomena: initial settlements, drainage and creep. The overconsolidated behaviour and complex compression characteristics of aged clays were not considered. Nevertheless, the model managed to produce a remarkably good fit to the measured settlements. Furthermore, its prediction of future settlements is consistent with those produced by more sophisticated models. This became possible because the model parameters were back-calculated using the field data. This example demonstrates the importance of simple geotechnical inverse analysis in understanding and predicting the settlements in large-scale land reclamation problems.

2.4 Mitigation Measures

Mitigation of damage due to global excessive settlements in land reclamation is a difficult task. The addition of a fill to compensate for these settlements after an island is complete is often not feasible technologically (e.g., the barges cannot reach the inside of the island) or due to operational reasons (it is unthinkable to shut down an airport). In addition, this fill, due to its weight, would probably cause additional settlements. Therefore, the negative effects of the settlements of the island on its ability to withstand high tides should be mitigated by extending the height of the seawalls accompanied by their reinforcement.

The major sources of worry, however, are the differential settlements affecting the structures. The passenger terminal building (Fig. 2.14a) is a structure with a key service floor area of 0.3 km^2. It consists of a 4 (3 + 1 basement) stories main building (320×150 m) and two 3 stories wing buildings (670×40 m each). These buildings are supported by 874 columns over their total length of 1,660 m.

The major problem with the main building is that, because of the basement, its weight represents only half of the weight of the soil it displaced. Therefore, to ensure that the island and the structure sank at the same speed rate, the basement of the terminal was lined with a quarter of a million tons of iron ore. As is seen in Figure 2.14b, this measure had only limited success – the solid line shows that the ground under the main building sank by October 2003 considerably less than the ground under the wing buildings. Most probably, the weight compensation was only partial.

In order to compensate for continuing differential settlements during the operation of the building, the supporting columns have been supplied with a jack-up system allowing for the adjustment of their heights. These adjustments have been performed two or three times a year by jack-ups and inserting thin plates (Matsui *et al.*, 2003) to keep the differential settlements within the design limits: $\theta = 1/400$ for local distortion angle of the roof structure of the main frame and $\theta = 1/600$ for the roof structure of the wings. The dashed line in Figure 2.14b indicates the level of the columns corrected by the jack-ups. As is seen, in spite of the fact that by October 2003 the maximum correction height has reached for some columns almost 60 cm, the differential settlement criteria had not been yet satisfied. The maximum differential settlement between the center of the main building and the lowest columns of the south wing building was 95 cm, resulting in the maximum inclination of the dashed line of $\theta = 1/450$, which exceeds by far the design limit of $\theta = 1/600$ for local distortion angle of the roof structure of the wings.

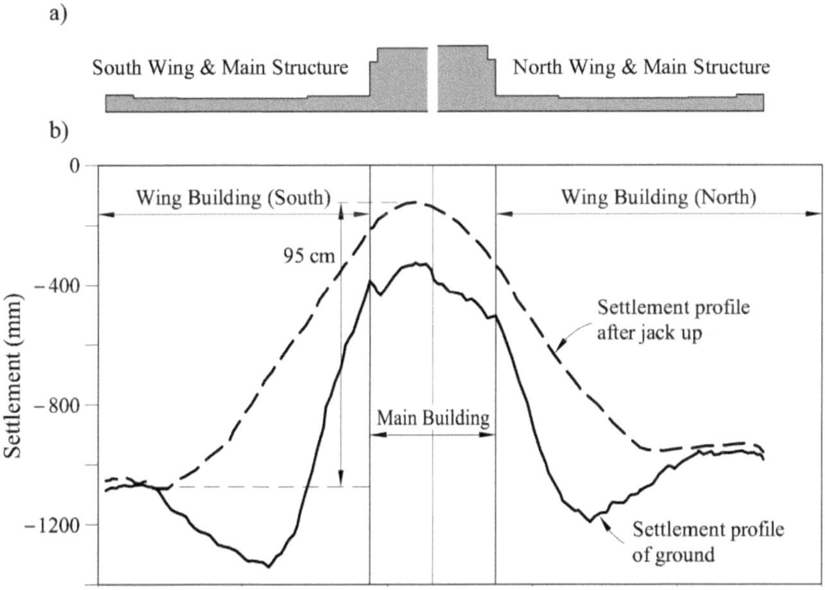

Figure 2.14 The passenger terminal building (after Akai and Tanaka, 2005: © 2005, IOS Press, used with permission): (a) outline; (b) settlement profiles along the longitudinal line.

2.5 Lessons Learned

2.5.1 High level of indeterminacy

The major problem with large-scale land reclamation projects is that it is almost impossible to provide an accurate prediction of the rates of settlement based solely on the results of site investigation and laboratory consolidation tests. The major reasons for that are large spatial variability of soil properties and drainage geometry and the fact that laboratory tests often produce values of the coefficients of consolidation and secondary compression within two orders of magnitude (i.e. 10^2) from the field values – with the field exhibiting more pervious behaviour. Therefore, such estimates can only be used as initial conditions for design.

2.5.2 Immediate settlements

Immediate settlements in saturated clay layers in a one-dimensional problem should be zero. In reality, however, there are always some immediate settlements caused by the three-dimensionality of the real problem, which involves non-negligible lateral and, hence, vertical strains. In the layered sand-clay systems, however, some immediate settlements occur even in a one-dimensional problem. This is due to the compressibility of the sand layers, where pores are sufficiently large to allow for almost immediate dissipation of pore water pressures.

2.5.3 Limited drainage

The rate of consolidation is determined to a large extent by the length of the drainage path. Conventionally, a sand layer between two clay layers is considered to work as a drain. There are cases, however, where some of the sand layers appear to be lenses, entirely enclosed within a clay layer. It is important to observe dissipation of the pore water pressures in sand layers during the construction, in order to determine their draining ability.

2.5.4 Secondary compression

It is important to remember that the consolidation process is not the only one controlling the rate of the settlements in clay. Even when the excess pore water pressure has completely dissipated, different physical phenomena cause continued settlement. While this secondary compression is present from the beginning of the consolidation process, towards the end of primary consolidation its contribution becomes more visible. The secondary compression may produce a significant increase in settlements long after the primary consolidation is over.

2.5.5 The observational method

Because of the high level of indeterminacy in such projects, the observational method should be adopted in design and construction. In this method, the design is left flexible to accommodate changes as construction proceeds. The changes should be based on the continuous monitoring of significant field parameters and on inverse analysis of the field measurements. This back calculation allows for model parameters to be updated using real field data and then utilized for the next stages of design.

References

Akai, K., Nakaseko, K., Matsui, T., Kamon, K., Sugano, M., Tanaka, Y. and Suwa, S. (1995) Geotechnical and geological studies on seabed Osaka Bay. *Proceedings of the 11th European Conference on Soil Mechanics and Foundation Engineering* 8, 8.1 – 8.6.

Akai, K. and Tanaka, Y. (1999) Settlement behaviour of an off-shore airport KIA. *Geotechnical Engineering for Transportation Infrastructure*, 1041–1046.

Akai, K. and Tanaka, Y. (2005) Ex-Post-Facto estimate of performance at the offshore reclamation of airport Osaka/KIA. *Proceedings of the 16th International Conference on Soil Mechanics and Geotechnical Engineering* 2, 1011 – 1014.

Endo, H., Oikawa, K., Komatsu, A. and Kobayashi, M. (1991) Settlement of diluvial clay layers caused by a large scale man-made island. *Geo-Coast 91'*, 177 – 182.

Handy, R.L. (2002) First-order rate equations in geotechnical engineering. *ASCE Journal of Geotechnical and Geoenvironmental Engineering* 128 (5), 416 – 425.

KALD (2009) Kansai Airport Land Development Company, Website: www.kald.co.jp

Matsui, T., Oda, K. and Tabata, T. (2003) Structures on and within man-made deposits – Kansai Airport. *Proceedings of the 13th European Conference on Soil Mechanics and Geotechnical Engineering* 3, 315 – 328.

Mesri, G. and Varhanabhuti, B. (2005) Secondary compression. *ASCE Journal of Geotechnical and Geoenvironmental Engineering* 131 (3), 398 – 401.

Terzaghi, K. (1943) *Theoretical Soil Mechanics*, John Wiley and Sons, New York.

Chapter 3

Leaning Instability: The Tower of Pisa, Italy

TABLE OF CONTENTS

A.M. Puzrin et al., *Geomechanics of Failures*, DOI 10.1007/978-90-481-3531-8_3,

Chapter 3

Leaning Instability: The Tower of Pisa, Italy

3.1 Case Description

The Tuscan town of Pisa became a powerful Mediterranean republic and a flourishing commercial center in the 11th to 13th centuries. After a decisive victory in a sea battle at Palermo against the Saracens in 1063, its colonies included Sardinia, Corsica, Elba, parts of Southern Spain, and Carthage. It reached the peak of its power in the 12th century as a naval base for the first Crusade to the Holy Land, but in the 13th century, a number of bad political choices and military defeats lead to its demise and domination by Genoa and Florence. Though being the birthplace of Galileo Galilei (1564 – 1642), the town is probably best known for its Romanesque Leaning Tower (Fig. 3.1a). It was built as a bell tower of the cathedral complex on Piazza dei Miracoli (the Square of Miracles, Fig. 3.1b). The construction of the complex was initially funded from the treasure found on six large, heavily loaded ships captured in 1063 from the Saracens, and was meant to demonstrate the power of the Church and the Republic. Excessive inclination, however, turned the Leaning Tower very early into a curiosity and a tourist attraction. In 1990, however, due to the alarming tilt of almost 5.5 degrees and recent collapse of the civic tower in Pavia, which caused four fatalities, it was closed to the public. An elaborate and expensive stabilization campaign followed, which returned the Leaning Tower to its inclination of 1844. In December 2001, it was again opened to the public.

(a) (b)

Figure 3.1 The Leaning Tower of Pisa (after Burland *et al.*, 1998): (a) the Leaning Tower; (b) Piazza dei Miracoli.

3.1.1 Construction

Construction of the Leaning Tower began in 1173, about 10 years after the completion of the Cathedral although the construction work for the Baptistery started about 20 years earlier.

The footprints of both the Cathedral and the Baptistery are much larger than that of the Leaning Tower and they do not experience any differential settlements.

The 56 m high Leaning Tower with a ground floor, six galleries, and a bell chamber is constructed as a hollow cylinder (Fig. 3.2). Its external and internal diameters at the ground floor are 15.5 and 7.4 m, respectively. The outer and inner walls are faced with high-quality white San Giuliano marble, while the cavity between them is filled with a typical rubble and mortar mixture. The tower is based on a shallow ring foundation, with an external diameter of 19.6 m and a width of 7.5 m.

After starting in 1173, the construction process lasted almost 200 years, and included three stages, with almost century-long breaks between them. The first stage lasted for about five years.

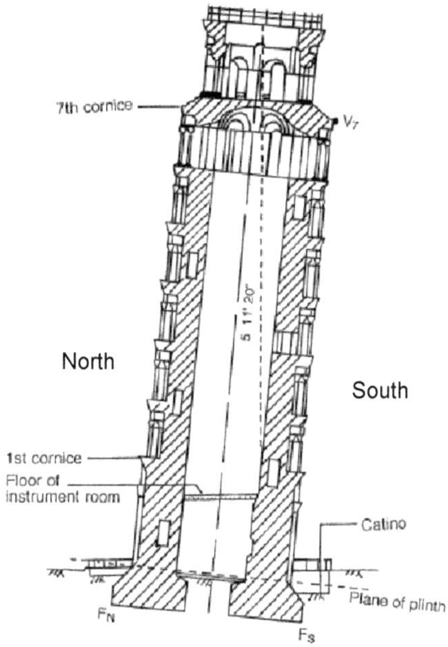

Figure 3.2 Cross-section of the Leaning Tower (after Burland *et al.*, 1998).

During the construction of the third gallery (the fourth level), the work had to be suspended, most probably due to the financial crisis caused by the war with Florence. After this first stage, the Leaning Tower was already leaning slightly to the north.

The second stage began around 1272, and by 1278 all six galleries (seven levels) were completed. At that time, the tower was already leaning to the south. Attempts to correct this tilt during the construction, by the shifting upper floors relative to the lower ones and building them vertically, resulted in a "banana" shape of the tower, but were only partially successful.

The third stage began in 1360 with the aim of building the bell chamber (the eighth level). By that time, the southward tilt of the tower was so significant that it was necessary to construct six steps on the south side compared to only four on the north side between the sixth gallery and the floor of the bell chamber. The building was completed in 1370.

3.1.2 The history of tilting

The builders tried to correct the tilt during its construction by having additional steps or stone wedges on the lower side of the tower. This allowed for the history of its inclination to be deduced from careful measurements of the relative inclinations of the masonry courses (Burland and Viggiani, 1994). From Stage I of the construction, when approximately 9,000 tons of the total 14,500 ton structure was built, the tower developed a tilt to the north, which by the beginning of Stage II in 1272 had reached 0.2 degrees (Fig. 3.3a). By the end of Stage II in 1278, however, when the load reached about 13,600 tons, it was already leaning to the south at an angle of 0.6 degrees. During a 90 year long break between the second and the third stages, the Leaning Tower kept tilting further south, so that by the beginning of Stage III in 1360, the inclination had reached 1.6 degrees.

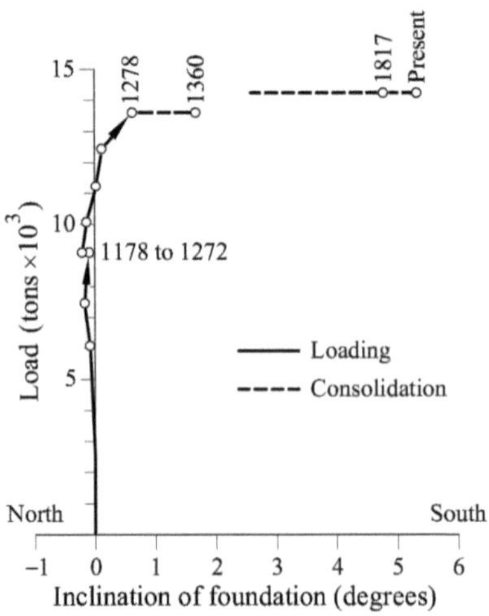

(a)

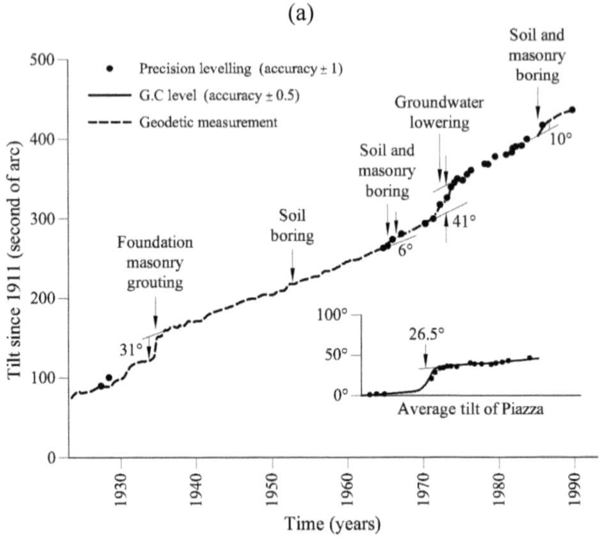

(b)

Figure 3.3 History of tilting (after Burland *et al.*, 1998): (a) as a function of load; (b) the modern measurements relative to 1911 (° in the figure denoting seconds of arc).

The first available measurement after construction was made in 1817 and the recorded tower inclination was about 4.8 degrees. It increased sharply to 5.3 degrees in 1838 when, in order to expose the columns of the ground floor, the architect Alessandro Della Gherardesca dug a walkway (catino) around the foundations and groundwater filled the excavation.

Since 1911, theodolite measurements have been made on a regular basis and in 1934, a pendulum was suspended from the sixth floor. It can be seen in Figure 3.3b that the inclination is increasing with time. In 1990, it reached 5.44 degrees and was growing by about 6 arc seconds per year (arc second is a unit of angular measure equal to 1/60 of an arc minute, or 1/3,600 of a degree).

The tower appears to be very sensitive to ground disturbances (such as the soil and masonry borings in 1934, 1966, and 1985) and to changes in the groundwater conditions (such as pumping from sand layers and lowering the groundwater level in the 1970s). The lowering of the groundwater level also caused a tilt of the entire Piazza (see the detail in Fig. 3.3b), although to a much smaller extent than that of the tower.

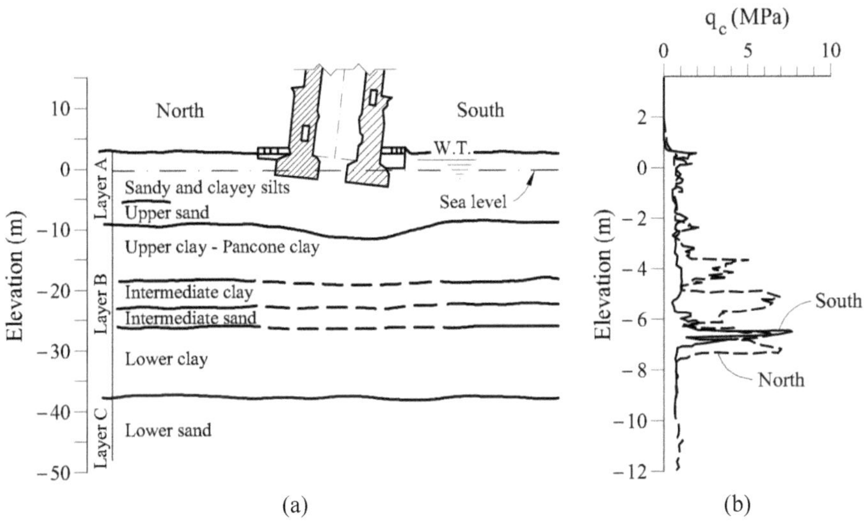

Figure 3.4 The ground conditions (after Burland *et al.*, 1998): (a) soil profile; (b) cone resistance profiles to the north and to the south of the Leaning Tower.

3.1.3 The problem

The soil profile below the Leaning Tower is represented by alternating sand and clay layers (Fig. 3.4a). The groundwater level (W.T. in Fig. 3.4a) is 1 to 2 m below the surface. A number of different factors may be responsible for the tilt of the tower. One of these factors – the foundation failure – has to be excluded due to

- the very slow tilt rate (should be much faster at failure);
- the local depression in the surface of the upper clay layer (Fig 3.4a) – an indicator of the volume change due to consolidation;

- the lack of heave in the ground surface at the southern side.

Another possible factor – differential settlements due to variability of soil stiffness (e.g., a locally thinner sand layer on the southern side, Fig. 3.4b) – can only be partially responsible for the tilt. (Recall that the Cathedral and the Baptistery do not experience any significant differential settlements.)

This leaves us with the main factor – the "leaning instability" – which apparently caused the rapid increase in the inclination of the tower towards the end of its construction (Burland and Potts, 1994).

3.1.4 The leaning instability

Leaning instability of a tall, narrow structure occurs at a critical height to width ratio when the overturning moment caused by a small inclination cannot be compensated by the corresponding resisting moment mobilized by the foundations (Hambly, 1985; 1990). This effect is well known to anyone who has attempted building Lego towers on a soft carpet. The leaning instability is due to the high soil compressibility and not to its low strength.

Note that both the Cathedral and the Baptistery have a much smaller height to width ratio than the Leaning Tower. While the leaning instability of the Pisa Tower has been extensively studied in the literature using sophisticated finite element analysis, which provided a strong basis for developing the strategy for its stabilization (e.g., Burland and Potts, 1994), the purpose of this chapter is to present a simplified model dealing with this phenomenon.

3.2 The Theory

3.2.1 Model assumptions

Consider a rigid structure with a rectangular plan $B \times L$ on two rectangular footings $b \times L$ based on a Winkler foundation (built of individual linear elastic springs) with the coefficient of subgrade reaction k (Figs. 3.5a,b). The contact pressure $q = k \cdot d$, where d is the settlement and the foundation reaction is $F = q \cdot L \cdot b = K \cdot d$ (Fig. 3.5c), where

$$K = L \cdot b \cdot k \tag{3.1}$$

is the spring constant. For simplicity, rotational angular stiffness is assumed to be zero, but the spring stiffness *K* is lower for loading than for unloading: $K_L < K_U$. In agreement with the Winkler assumption, there is no interaction between the two foundations and no horizontal resistance.

3.2.2 Equivalent foundations

For other shapes of structures and foundations, e.g. for the most common ring and hollow square foundations in Figure 3.6, an equivalent pair of rectangular foundations of the type shown in Figure 3.5b can be defined, provided the area and the moment of inertia of the two foundations are identical. For a ring foundation with the average radius r and the width b (Fig. 3.6a), its area A_r, and the inertia moment I_r with respect to its axis of symmetry are

$$A_r = 2\cdot\pi\cdot r\cdot b, \quad I_r = \frac{\pi}{4}\cdot r\cdot b\cdot\left(4r^2+b^2\right). \tag{3.2}$$

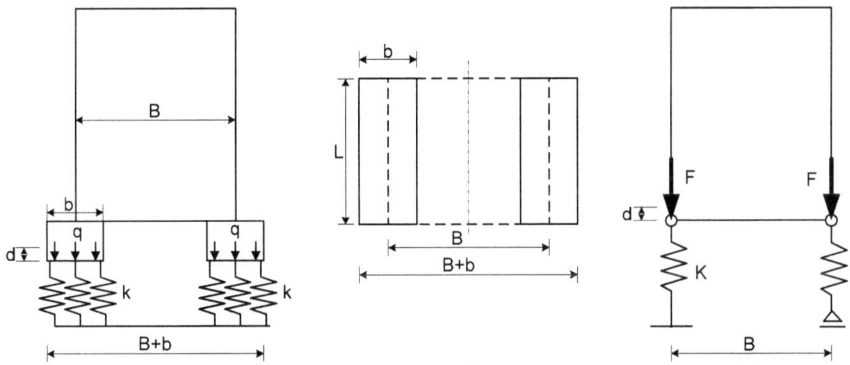

(a) (b) (c)

Figure 3.5 The simplified model: (a) structure and the soil; (b) plan view of foundations; (c) resultant foundation reactions.

The corresponding parameters for the equivalent pair of rectangular foundations, with the same width b (Fig. 3.5b) are

$$A_{eq} = 2\cdot L\cdot b, \qquad I_{eq} = \frac{L\cdot b^3}{6}+\frac{B^2}{2}L\cdot b. \tag{3.3}$$

Equating expressions (3.2) and (3.3), we obtain the geometry of the equivalent foundation:

$$L=\pi r, \qquad B=\sqrt{2r^2+b^2/6}. \tag{3.4}$$

For a hollow square foundation with average half side r and width b (Fig. 3.6b), its area A_S and the inertia moment I_S with respect to its axis of symmetry are

$$A_s = 8\cdot r\cdot b, \qquad I_s = \frac{4}{3}\cdot r\cdot b\cdot\left(4r^2+b^2\right). \tag{3.5}$$

Equating expressions (3.3) and (3.5), we obtain geometry of the equivalent foundation:

$$L=4\cdot r, \qquad B=\sqrt{8r^2/3+b^2/3}. \tag{3.6}$$

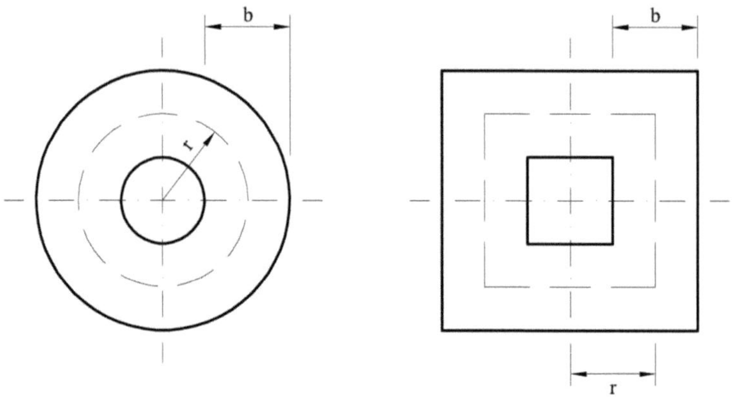

Figure 3.6 The equivalent foundations: (a) ring; (b) hollow square.

3.2.3 Overturning moment due to an incremental inclination

In the vertical position, both foundation springs would be loaded by equal loads $F = G/2$, where G is the weight of the structure, which would not produce any differential settlements. Inclination by an angle α causes a horizontal shift in the position of its center of gravity (Fig. 3.7a)

$$x = H_c \cdot \sin\alpha, \tag{3.7}$$

and a relative differential settlement

$$\frac{d_L - d_U}{B} = \sin\alpha, \tag{3.8}$$

Where H_c is the height of the center of gravity. As a result (Fig. 3.7b), the force in the spring in the direction of inclination will increase to F_L (loading), while the force in the opposite spring will decrease to F_U (unloading). From the equilibrium of the vertical forces it follows that

$$F_U + F_L = G. \tag{3.9}$$

In the next section, it will be shown that the center of rotation O is located at the bottom of the structure at some distance y from the axis of symmetry. Stability of the structure against overturning requires that the overturning moment with respect to O,

$$M_O^{\text{over}} = G \cdot (x + y\cos\alpha) = G \cdot H_c \cdot \sin\alpha + G \cdot y \cdot \cos\alpha, \tag{3.10}$$

does not exceed the resisting moment

$$M_O^{\text{res}} = (F_L - F_U) \cdot \frac{B}{2} \cdot \cos\alpha + (F_L + F_U) \cdot y \cdot \cos\alpha. \tag{3.11}$$

From the equilibrium Equation (3.9) it follows that the second terms in expressions (3.10) and (3.11) are identical, and therefore do not affect the stability so that the stability condition can be written as

$$M_O^{\text{over}} \leq M_O^{\text{res}} \quad \Rightarrow \quad \frac{H_c}{B} \leq \frac{\Delta F}{G} \cot \alpha, \tag{3.12}$$

where

$$\Delta F = \frac{F_L - F_U}{2} = F_L - \frac{G}{2} = \frac{G}{2} - F_U. \tag{3.13}$$

The question now is: can the foundations mobilize a sufficiently large force difference ΔF to ensure stability against the overturning?

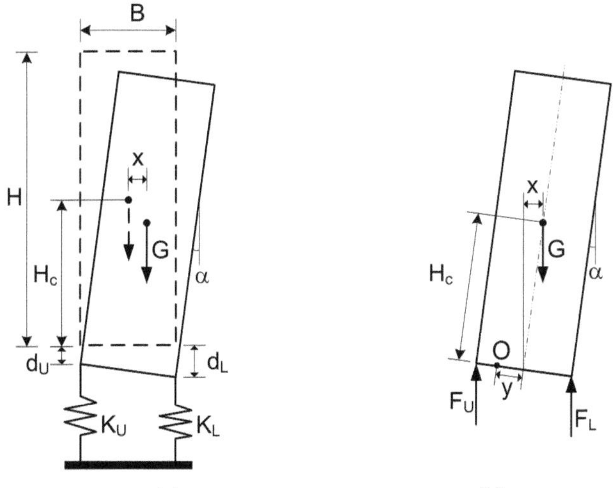

Figure 3.7 Incremental inclination: (a) geometry; (b) overturning moment.

3.2.4 Resisting moment mobilized by the foundations

In the vertical position, both foundation springs are loaded by $F = G/2$, and experience the settlement of d_G, whose value depends on the loading history (Fig. 3.8a). Once the structure rotates, the foundations are loaded and unloaded by a force increment ΔF, and the foundation settlements become d_L and d_U, respectively (Fig. 3.8b). The foundation stiffness in loading is smaller than that in unloading: $K_L < K_U$. From Figure 3.8a, it follows that

$$d_L - d_U = \frac{\Delta F}{K_L} + \frac{\Delta F}{K_U}, \tag{3.14}$$

which, using Equation (3.8), gives the force difference

$$\Delta F = \frac{K_L K_U}{K_L + K_U} B \sin\alpha \tag{3.15}$$

and the position of the center of rotation O (Fig. 3.8b)

$$y = \frac{B}{2} \cdot \frac{K_U - K_L}{K_U + K_L}. \tag{3.16}$$

Substitution of Equation (3.15) into the stability condition (3.12) gives

$$\frac{H_c}{B} \le \frac{B}{G} \cdot \frac{K_L \cdot K_U}{K_L + K_U} \cdot \cos\alpha. \tag{3.17}$$

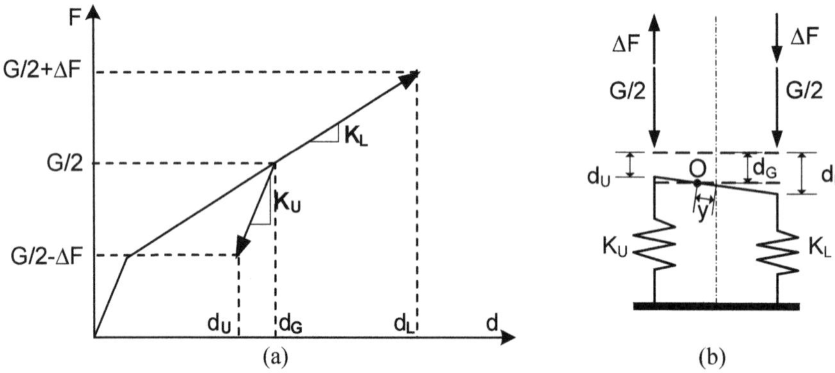

Figure 3.8 Resisting moment mobilized by foundations: (a) load-displacement foundation behaviour; (b) kinematics.

3.2.5 Spring coefficients

The foundations spring constants K_L and K_U, unless derived from a full-scale field test of the type presented in Figure 3.8a, can be estimated using the following considerations. A conservative stability criterion requires a lower estimate for the right side of inequality (3.17). This, in turn, requires lower estimates for the values of foundations spring constants K_L and K_U. First of all, it has been shown (Cheney *et al.*, 1991) that the Winkler theory is more conservative than the elastic half-space theory due to the difference in stress distribution beneath the tower. Secondly, a spring constant is related through Equation (3.1) to the coefficient of subgrade reaction k limited by the following expression (Lang *et al.*, 2007):

$$k \ge \frac{M_E}{f \cdot b}, \tag{3.18}$$

where M_E is the compression modulus; f is a shape factor derived from the plot in Figure 3.9a. Equality in (3.18) is achieved for an infinitely thick compressible layer: finite thickness leads to a stiffer response, i.e. a larger k.

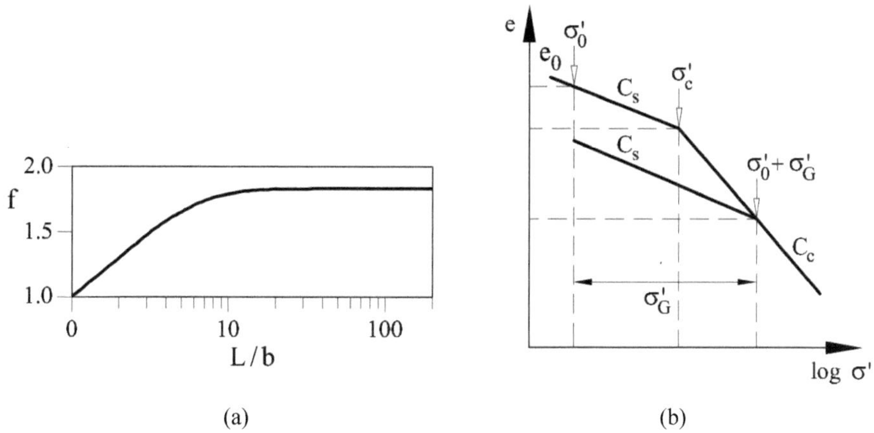

Figure 3.9 Soil stiffness in loading and unloading: (a) correction factor for the coefficient of subgrade reaction (after Lang *et al.*, 2007); (b) a laboratory consolidation test.

Thirdly, for clays each spring can be presented as an individual oedometer, and the tangent compression modulus M_E in loading and unloading can be derived from a consolidation test in Figure 3.9b (tangent modulus corresponds to the small variation in inclination)

$$M_{EL} = \ln 10 \cdot (1+e_0)\frac{\sigma'_0 + \sigma'_G}{C_c}, \qquad M_{EU} = \ln 10 \cdot (1+e_0)\frac{\sigma'_0 + \sigma'_G}{C_s}, \quad (3.19)$$

where $\sigma'_G = G/2Lb$ is the contact foundation pressure for the vertical structure;
C_c is the compression index;
C_s is the swelling index;
e_0 is the in situ void ratio;
σ'_0 is the initial stress;
$\ln 10 \approx 2.30$.

The lower estimates for expressions (3.19) are then given by ignoring the initial stresses:

$$M_{EL} > 2.3 \cdot (1+e_0)\frac{\sigma'_G}{C_c}, \qquad M_{EU} > 2.3 \cdot (1+e_0)\frac{\sigma'_G}{C_s}. \quad (3.20)$$

Finally, substitution of inequalities (3.18) and (3.20) into (3.1) produces the lower bounds for the spring constants:

$$K_L > 2.3 \cdot (1+e_0)\frac{G}{2 \cdot f \cdot b \cdot C_c}, \qquad K_U > 2.3 \cdot (1+e_0)\frac{G}{2 \cdot f \cdot b \cdot C_s}. \quad (3.21)$$

3.2.6 Criteria for leaning instability

In the case when the values of foundation spring constants are known from a full-scale field test, the criterion for the stability of a vertical structure is obtained from inequality (3.17) with $\alpha = 0$:

$$\frac{H_c}{B} \le \frac{B}{G} \frac{K_L K_U}{K_L + K_U}. \tag{3.22}$$

In most cases, however, the full-scale foundation test is not possible and the stability criterion has to be based on the laboratory estimates of the soil stiffness. This criterion is obtained by substitution of expression (3.21) into (3.22):

$$\frac{H_c}{B} \le \frac{B}{f(L/b)\ b} \cdot \frac{1.15 \cdot (1 + e_0)}{C_c + C_s}, \tag{3.23}$$

where f is a form factor derived from the plot in Figure 3.9a. If this inequality is not satisfied, a small deviation from the vertical position will cause the structure to fall.

For a ring foundation, substituting Equation (3.4) into (3.23) gives

$$\frac{H_c}{r} \le \frac{\rho^2 + 1/12}{f(\pi \cdot \rho) \cdot \rho} \cdot \frac{2.3 \cdot (1 + e_0)}{C_c + C_s}, \tag{3.24}$$

where

$$\rho = r/b, \qquad \rho \ge \sqrt{5/12} \approx 0.65. \tag{3.25}$$

The inequality in (3.25) follows from the condition $B \ge b$ applied to the second Equation (3.4), so that the two equivalent foundations in Figure 3.5b would not overlap.

For a hollow square foundation, substituting Equations (3.6) into (3.23) gives

$$\frac{H_c}{r} \le \frac{4 \cdot \rho^2 + 1/2}{3 \cdot f(4 \cdot \rho) \cdot \rho} \cdot \frac{2.3 \cdot (1 + e_0)}{C_c + C_s}, \tag{3.26}$$

where

$$\rho = r/b, \qquad \rho \ge 0.5. \tag{3.27}$$

The inequality in (3.27) follows from the condition $B \ge b$ applied to the second Equation (3.6). Using criterion (3.24), we can attempt to define a minimum H_c/r ratio, for which the above leaning instability analysis becomes meaningful. The minimum to the right side of inequality (3.24) is achieved at $\rho = \sqrt{5/12} \approx 0.65$, for which $f(\pi\rho) \approx 1.3$. Because for many clays $(1 + e_0)/(C_c + C_s) \ge 2$, any ratio higher than

$$H_c/r = 2.74 \tag{3.28}$$

should in principle be checked using criterion (3.24). In simple words, any tower on a shallow ring foundation with a height larger than about 3 diameters may be subjected to the leaning instability problems!

For a hollow square foundation, similar arguments provide $H_c/r = 3.54$, indicating significantly higher efficiency of a hollow square foundation with respect to avoiding the leaning instability compared to a ring foundation.

3.2.7 Safety factors

Equations (3.22) – (3.27) provide the critical height to width ratios for the design of new structures. For an existing structure already inclined by angle α, the following corresponding definitions of the safety factors against the leaning instability may be suggested, based on the ratio between the critical (see Eq. (3.17)) and existing H_c/B ratios:

$$F_s = \frac{(H_c/B)_{cr}}{H_c/B} = \frac{B^2}{H_c \cdot G} \cdot \frac{K_L \cdot K_U}{K_L + K_U} \cos\alpha \tag{3.29}$$

or

$$F_s = \frac{B^2}{f(L/b) \cdot H_c \cdot b} \cdot \frac{1.15 \cdot (1 + e_0)}{C_c + C_s} \cdot \cos\alpha. \tag{3.30}$$

For a ring foundation this becomes

$$F_s = \frac{r}{H_c} \cdot \frac{\rho^2 + 1/12}{f(\pi\rho)\ \rho} \cdot \frac{2.3 \cdot (1 + e_0)}{C_c + C_s} \cos\alpha, \tag{3.31}$$

and, for a hollow square foundation, we obtain

$$F_s = \frac{r}{H_c} \cdot \frac{4 \cdot \rho^2 + 1/2}{3 \cdot f(4 \cdot \rho) \cdot \rho} \cdot \frac{2.3 \cdot (1 + e_0)}{C_c + C_s} \cdot \cos\alpha. \tag{3.32}$$

3.2.8 Bearing capacity

Inclination of the structure (Fig. 3.7) increases the load F_L on one of the footings (the other one is being unloaded). This load and the corresponding contact stress σ_L can be found from the equilibrium of vertical forces (3.9) and moments (3.10) and (3.11):

$$F_L = \frac{G}{2} + \frac{GH_c}{B} \tan\alpha, \qquad \sigma_L(\alpha) = \frac{G}{b \cdot L}\left(\frac{1}{2} + \frac{H_c}{B} \tan\alpha\right). \tag{3.33}$$

Once the structure inclination starts increasing (e.g., due to the leaning instability) this contact stress may exceed the bearing capacity σ_f and the foundation will fail. In the absence of cohesion, this bearing capacity for drained loading is given by the following modification of Terzaghi formula (Lang *et al.*, 2007):

$$\sigma_f = (\gamma' \cdot t + q) \cdot N_q \cdot s_q \cdot d_q + \frac{1}{2} \cdot b \cdot \gamma' \cdot N_\gamma \cdot s_\gamma \cdot d_\gamma, \tag{3.34}$$

where γ' is the effective unit weight of soil; t is the foundation depth;

$$N_q = \exp(\pi \tan\varphi) \cdot \tan^2\left(45° + \frac{\varphi}{2}\right), \quad N_\gamma \cong 1.8 \cdot (N_q - 1) \cdot \tan\varphi \tag{3.35}$$

are the bearing capacity factors;

$$s_q = 1 + \frac{b}{L}\tan\varphi, \qquad s_\gamma = 1 - 0.4\frac{b}{L} \tag{3.36}$$

are the shape correction factors;

$$d_q = 1 + 0.035\tan\varphi(1 - \sin\varphi)^2 \arctan\left(\frac{t}{b}\right), \quad d_\gamma = 1 \tag{3.37}$$

are the depth correction factors.

The safety factor against bearing capacity failure can then be defined by the ratio between the bearing capacity pressure and the existing contact pressure of the loaded footing:

$$F_s(\alpha) = \frac{\sigma_f}{\sigma_L(\alpha)}. \tag{3.38}$$

The solution of Equation $F_s(\alpha_f) = 1$ will produce the value of the inclination angle α_f, at which the structure will fail.

3.2.9 Summary

The conclusions from the above analysis can be summarized as follows:

1. In addition to the regular bearing capacity and settlement calculations, any structure on shallow foundations with a height to diameter ratio larger than three should be checked for the possibility of a leaning instability failure.
2. This check can be performed using a simple formula.
3. The formula uses three geometric parameters: the height of the centre of gravity, as well as the average radius and the width of the foundation.
4. In addition, three soil parameters have to be found from laboratory tests: the in situ void ratio, as well as the compression and swelling indexes.
5. If the existing structure is already inclined, the probability of the leaning instability increases.
6. The safety factor against the leaning instability can be calculated using one additional parameter: the inclination of the structure.

3.3 The Analysis

The theory presented above provides tools for a simplified analysis of the stability of the Leaning Tower of Pisa.

3.3.1 Simplified model

The relevant tower geometry can be summarized as follows:

- total height of the tower: $H = 56.0$ m;
- height of the center of gravity: $H_c = 22.6$ m;
- width of the ring foundation: $b = 7.5$ m;
- average radius of the ring foundation: $r = 6.05$ m;
- initial foundation depth: $t_0 = 2.0$ m.

The corresponding loads are:

- total weight of the tower: $G = 142.5$ MN;
- average vertical stress: $\sigma = 500$ kPa.

The average inclination in the beginning of the 90s: $\alpha = 5.44°$.

The bearing capacity failure would most likely take place within the upper silt and sand layers, because the depth is larger than $b = 7.5$ m (Fig. 3.4a). The ground water level is assumed to be 1.5 m below the groundsurface. Due to relatively high permeability of these layers, the failure would take place under drained conditions. The corresponding effective soil properties recommended by Rampello and Callisto (1998) for these layers are summarized below:

- friction angle: $\varphi'_{cv} = 34°$;
- effective cohesion: $c' = 0\,\text{kPa}$;
- effective unit weight of soil: $\gamma' = 9\,\text{kN/m}^3$.

The leaning instability would most likely take place due to the consolidation settlements of the upper clay layer. The soil properties recommended by Rampello and Callisto (1998) for this clay layer (grey-blue silty clay of high plasticity $I_p = 30 - 50\%$) are summarized below:

- compression index: $C_c = 0.90$;
- swelling index: $C_s = 0.15$;
- in situ void ratio: $e_0 = 1.5$.

The first question to answer is: could this inclination be caused by a bearing capacity failure?

3.3.2 Bearing capacity

First, we calculate the bearing capacity of the tower close to the end of its construction when it was still standing straight. The bearing capacity calculation requires the length of the equivalent footing, which is calculated using the first Equation of (3.4):

$$L = \pi \cdot r = 3.14 \times 6.05 = 19.00\ \text{m}.$$

The width of the equivalent footing is $b = 7.5$ m and its depth $t = 2.0$ m. The friction angle is $\varphi'_{cv} = 34°$. Substitution of the above parameters into the formulas (3.35) – (3.37) gives

$$N_q = 29.4, \quad s_q = 1.27, \quad d_q = 1.07,$$

$$N_\gamma = 34.5, \quad s_\gamma = 0.84, \quad d_\gamma = 1.00.$$

Then the drained bearing capacity (3.34) can be calculated as

$$\sigma_f = (19 \times 1.5 + 9 \times 0.5) \times 29.4 \times 1.27 \times 1.07 + \frac{1}{2} \times 7.5 \times 9 \times 34.5 \times 0.84 \times 1.00 = 2{,}300 \text{ kPa}. \tag{3.39}$$

Then the safety factor against the bearing capacity failure for a not inclined Tower of Pisa was

$$F_s = \frac{\sigma_f}{\sigma} = \frac{2{,}300}{500} = 4.6. \tag{3.40}$$

This is more than sufficient and could, in principle, justify the choice of the shallow foundation for the tower construction.

Now, when the tower is inclined, the pressure under one of the footings is much higher than the average one. Calculating from the second Equation (3.4): $B = 9.1$ m, from the second Equation (3.33) we obtain

$$\sigma_L(\alpha) = \frac{142{,}500}{7.5 \times 19}\left(\frac{1}{2} + \frac{22.6}{9.1}\tan 5.44^\circ\right) = 750 \text{ kPa}. \tag{3.41}$$

Nevertheless, the bearing capacity factor is still sufficiently large:

$$F_s = \frac{\sigma_f}{\sigma_L} = \frac{2{,}300}{750} = 3.07 > 3. \tag{3.42}$$

It has to be mentioned that the above analysis is not conservative, because the true failure mechanism will most probably go deeper into the weaker upper clay layer. It can be shown, however, that the safety factor would still be larger than one. It is, therefore, unlikely that the bearing capacity failure was the source of the tower's inclination and we need to look for another reason. Since the ratio (3.28) is

$$H_c / r = 22.6/6.0 = 3.77 > 2.74, \tag{3.43}$$

it is worth exploring the possibility of the leaning instability.

3.3.3 Leaning instability

Substitution of the relevant parameters into inequality (3.24) gives the following stability condition (note that $\rho = 0.81$, so that $\pi\rho = 2.5$ and $f = 1.4$ from the plot in Fig. 3.9a):

$$\frac{H_c}{r} \leq 3.56. \tag{3.44}$$

Comparing this critical value with the tower ratio from Equation (3.24), we conclude that the Leaning Tower is unstable, with the safety factor against the leaning instability:

$$F_s = \frac{3.56}{3.77} \cos 5.44^\circ = 0.94 < 1.00. \tag{3.45}$$

Note that, unlike for the bearing capacity safety factor (3.42), the inclination of the tower has a rather negligible effect on the safety factor against the leaning instability ($\cos\left(5.44^\circ\right) = 0.9955$).

In fact, already by the end of the Stage II of the construction in 1278, when the Leaning Tower still stood almost vertical, reaching the height of 48 m and weight of 13,600 ton, its center of gravity reached the height of $H_c = 20.7$ m, so that the safety factor against the leaning instability became

$$F_s = \frac{3.56}{20.7/6.0} = 1.03, \tag{3.46}$$

i.e. very close to failure. It is, therefore, not surprising that the tower inclination rapidly accelerated exactly at this time!

3.3.4 Discussion

The simplified geotechnical model of the leaning instability of the Leaning Tower of Pisa is based on the Winkler hypothesis and equivalent foundation approach and, therefore, cannot account for the spatial variability of geometry, loading, and soil properties, as well as for the three-dimensional effects of stress and strain distribution. It also ignores the important issue of the development of inclination in time. Nevertheless, the model managed to provide a reasonably good prediction of the onset of the leaning instability by the end of the second stage of the tower construction. This prediction was not back-calculated! It is based on only six parameters – three from the tower geometry (which do not leave much room for an interpretation) and three from the standard laboratory consolidation tests performed on Pisa clays. These test results, of course, have some considerable scatter but we used the values recommended by Rampello and Calisto (1998), who performed this extensive experimental program. This example demonstrates how a simple geotechnical model can predict such a complex phenomenon as the leaning instability, provided it incorporates the most essential parameters controlling this phenomenon.

3.4 Mitigation Measures

The stabilization campaign for the Leaning Tower of Pisa is one of the most exciting pages in its long history. Following the collapse in 1989 of the civic tower in Pavia, the Italian government appointed a Commission to advise on the stability of the Pisa tower. The geotechnical expertise in this Commission was represented by Professors M. Jamiolkowski (chairman), J. Burland, G. Leonards, and C. Viggiani who came up with a stabilization plan which included three stages (Fig. 3.10).

(a)

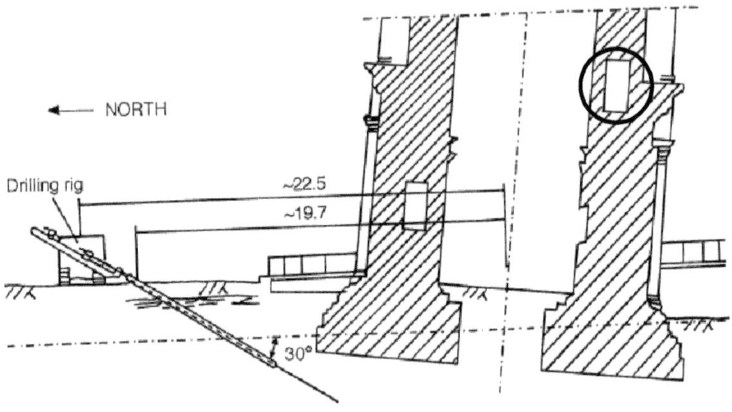

(b)

Figure 3.10 Geotechnical stabilization measures (after Burland *et al.*, 1998): (a) temporary; (b) permanent.

Stage I was a temporary structural reinforcement of the tower in the area of the maximum stress concentration in the wall, where the tower changes its wall thickness (circled in Fig. 3.10b). For this purpose, in 1992 temporary lightly pre-stressed plastic covered steel tendons were installed around the structure at the first cornice and at intervals up the second storey.

Stage II was a temporary reduction of the overturning moment. For this purpose, a temporary precast concrete ring was built in 1993 around the base of the tower and 690 tons of lead ingots were placed on its northern side (Fig. 3.10a). This reduced the tower inclination by almost one arc minute.

Stage III was a permanent reduction in the tower inclination. Originally this was supposed to be achieved by construction of ground anchors, which would replace the ingots by applying a controlled vertical load at the northern side. However, when in September of 1995 (to which the members of the Commission refer as the "black September"), the construction work started with an attempt to stabilize the surrounding soil by ground freezing, the tower tilted dangerously fast, and the operation had to be aborted. Therefore, an alternative technique – the soil extraction (Fig. 3.10b) – had to be adopted. This technique has been widely used in Mexico City and is described in Chapter 1. It was successfully applied to the Pisa tower and by May 2001, the tower went back by about half a degree, e.g. returned to its inclination of the 1840s. Finally, the catino and the tower foundations were grouted together. It would be, of course, possible to reduce the inclination even further, but nobody really wants to see a straight Tower of Pisa!

3.5 Lessons Learned

3.5.1 Leaning instability

For tall, narrow structures on soft soils, leaning instability can initiate the failure by increasing a small initial inclination caused by differential settlements.

3.5.2 Failure

Once the structure is inclined, it may eventually collapse due to the structural or foundation bearing capacity failure.

3.5.3 Deep foundations

If the leaning instability represents a potential problem, the use of shallow foundations should be avoided, unless they can be made sufficiently wide. Deep foundation (e.g., piles) could help to improve the stability of the structure by increasing the foundation stiffness; the tensile resistance of piles is in this case also beneficial.

3.5.4 Soil extraction

For existing inclined structures, the soil extraction has proven to be a reliable technique for their controlled stabilization.

References

Burland, J.B., Jamiolkowsky, M. and Viggiani, C. (1998) Stabilizing the leaning tower of Pisa. *Bulletin of Engineering Geology and the Environment* 57, 91 – 99.

Burland, J.B. and Potts, D.M. (1994) Development and application of a numerical model for the leaning tower of Pisa. *Proceedings of the First International Conference on Pre-failure Deformation Characteristics of Geomaterials,* Sapporo, Japan, 715 – 735.

Burland, J.B. and Viggiani, C. (1994) Observazioni sulcomportamento della Torre di Pisa. *Rivista Italiana di Geotecnica* 28, 179 – 200.

Cheney, J.A., Abghari, A. and Kutter, B.L. (1991). Stability of leaning towers. *ASCE Journal of Geotechnical Engineering* 117 (2), 297 – 318.

Hambly, E.C. (1985) Soil buckling and the leaning instability of tall structures. *The Structural Engineer* 63A (3), 77 – 85.

Hambly, E.C. (1990) Overturning instability. *ASCE Journal of Geotechnical Engineering* 116 (4), 704 – 709.

Lang, H.J., Huder, J., Amann, P. and Puzrin, A.M. (2007) *Bodenmechanik und Grundbau*. Springer Verlag, Berlin, 354 pp.

Rampello, S. and Callisto, L. (1998) A study on the subsoil of the Tower of Pisa based on results from standard and high-quality samples. *Canadian Geotechnical Journal* 35, 1074 – 1092.

PART II

BEARING CAPACITY

Chapter 4
Bearing Capacity Failure: Transcona Grain Elevator, Canada

TABLE OF CONTENTS

A.M. Puzrin et al., *Geomechanics of Failures*, DOI 10.1007/978-90-481-3531-8_4,

Chapter 4

Bearing Capacity Failure: Transcona Grain Elevator, Canada

4.1 Case Description

In September 1913, the Canadian Pacific Railway Company completed construction of a million-bushel (about 36,400 m^3) grain elevator at North Transcona, 11 km north-east of Winnipeg, Canada. The elevator was one of the most important structures and one of the largest gravity railroad yards in the world, which covered several square miles and was built on partly farmed, relatively flat prairie land. The purpose of the elevator was to provide relief for the Winnipeg Yards during the months of peak grain-shipment.

The structure consisted of a reinforced-concrete work-house, and an adjoining bin-house, which contained five rows of 13 bins, each 28 m in height and 4.4 m in diameter. The bins were based on a concrete structure containing belt conveyors supported by a reinforced-concrete shallow raft foundation.

After the structure was completed, the filling was begun and grain was distributed uniformly between the bins. On October 18, 1913, after the elevator was loaded to 87.5% of its capacity, settlement of the bin-house was noted. Within an hour, the settlement had increased uniformly to about 30 cm following by a tilt towards the west (Fig. 4.1a,b), which continued for almost 24 hours until it reached an inclination of almost 27 degrees (Allaire, 1916).

(a) (b)

Figure 4.1 The collapse of the Transcona Grain Elevator: (a) view looking southwest (Engineering News, 1913); (b) view looking northeast (White, 1953: © 1953 Thomas Telford Limited. Used with permission).

Many years later, when soil mechanics had provided the basis for computing the ultimate bearing-capacity of soils, it was realized that the Transcona failure afforded one of the best of the few opportunities for a full-scale check on the validity of its assumptions (Peck and Bryant, 1953).

4.1.1 Construction

Excavation for the elevator foundations started in 1911. The first 1.5 m of the ground were rather soft, but then the excavation penetrated into a relatively stiff blue clay, typical for that area, and locally known as the "blue gumbo". No borings were taken (White, 1953), but after the excavation reached its design depth of 3.7 m, field-bearing capacity tests were carried out by loading a plate laid upon a prepared smooth clay surface. The test loading was applied using a specially constructed wooden framework.

The tests indicated that the soil was capable of bearing a uniformly distributed load of at least 400 kPa (Engineering News, 1913; Allaire, 1916). Because the maximum foundation pressure was not supposed to exceed 300 kPa, the tests appeared to satisfy the requirements of the engineers. They assumed that the "blue gumbo" at the site had similar characteristics and a depth to that on which similar raft foundations of many heavy structures had been founded in the vicinity of Winnipeg.

A 60 cm thick 23.5×59.5 m reinforced-concrete slab (Fig. 4.2a) was built to serve as a foundation for the concrete framework of the underground conveyor belt tunnels supporting the bin-house. The construction of the bin-house proceeded at a rapid rate during the autumn and winter of 1912, the concrete circular bins being raised at the rate of 1 m per day, until they reached the designed height of 28 m. The 55 m high work-house was built about 3 m to the south of the bin-house, on a 21.5×29.3 m slab foundation (Fig. 4.2a). The only connection between the two buildings was a bridge for a conveyor belt, operating in a low cupola at the top of the bin-house. In September 1913, the construction of the elevator was completed (Fig. 4.2b), and its filling with grain was begun. At that stage the structures stayed straight and no excessive settlements were observed.

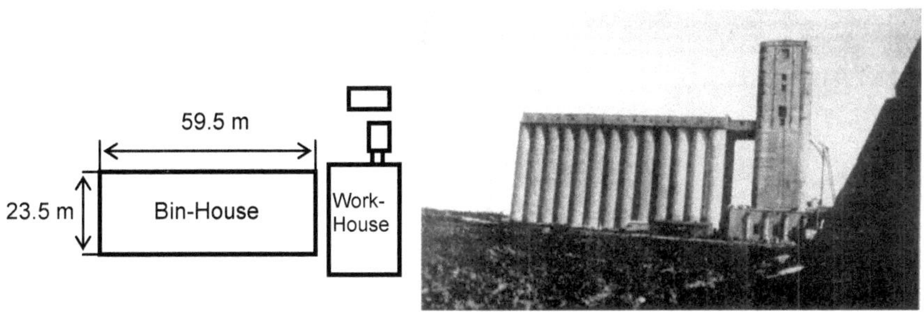

(a) (b)

Figure 4.2 The Elevator: (a) a foundation plan; (b) completed, before the collapse (after White, 1953: © 1953 Thomas Telford Limited. Used with permission).

4.1.2 The failure

The first troubles started about half a year before the end of the construction, though they were not directly related to the elevator. In the spring of 1913, after the melting of heavy winter snows, a great deal of trouble was caused throughout the railroad yards by large settlements of the embankments carrying the tracks (White, 1953). In one case, high-level tracks, which were laid on a 9 m high embankment, subsided by more than a meter, causing a significant heave of subsoil on the sides of the embankment.

The major incident happened on Saturday, 18th October, 1913. Between 11 and 12 a.m., with the elevator containing 875,099 bushels of grain, movement was noticed on the bridges between the bin- and work-houses. By 1 p.m., the bin-house had already settled about 30 cm. The 7.5 – 9.0 m wide strip of ground around the bin-house (except for the south side, where the work-house stood) heaved up 1.2 – 1.5 m. In the afternoon, the settlement rate became higher on the west side, producing a tilt to the west.

The structure continued settling and tilting until noon Sunday, 19 October. However, as the structure tilted to the west, the earth on that side bulged up, forming a cushion which slowed down the movement. The movement of the bin structure was gradual and barely susceptible to the eye, but a considerable amount of commotion was caused by the connecting bridges carrying the conveyor belts breaking down and crashing to the ground. During the night, the cupola structure housing the conveyor over the bins suddenly collapsed and fell to the ground. This reduced the load and there was subsequently little further movement (White, 1953). The final position of the bin-house was at an angle of about 27 degrees from the vertical, with the east side raised 1.5 m from its original level, opening a gap in the ground, while the west side was 9 m below its normal level.

The main concern of everyone at the time was whether the work-house could stand the disturbance. Check levels were therefore taken and it was found to be standing firm. Naturally a great strain was put on the comparatively thin walls of the storage bins, but they were apparently significantly over-designed so that the bin-house showed hardly any damage, apart from a few small hair cracks (Engineering News, 1913).

4.1.3 The problem

Several wash-borings were made immediately after the failure, showing that the elevator was underlain by rather uniform deposits of clay. This finding was in agreement with the geological history of the area, according to which extensive fine-grained sediments were deposited in the waters of the glacial Lake Agassiz which came into being when the Wisconsin ice-sheet blocked the region's northern outlet. Winnipeg lies above one of the deeper portions of the lake basin and, as a consequence, about 9 to 17 m of laminated sediments are found overlying the Ordovician limestone bedrock.

The wash-borings, therefore, confirmed the designers' assumptions, and the failure of the Transcona Grain Elevator remained a mystery for another 40 years. Indeed, if the smaller-scale plate loading tests predicted a safety factor of more than 1.3, and the soil profile is homogeneous, how could the foundation fail?

The answer to this question was given by Peck and Bryant (1953) who, in 1951, made two additional borings, far enough from the zone of failure to be in material unaffected by the displacements. They obtained undisturbed soil samples and performed unconfined compression strength tests (triaxial shear tests with zero confining stress), which produced some eye-opening results (Fig. 4.3).

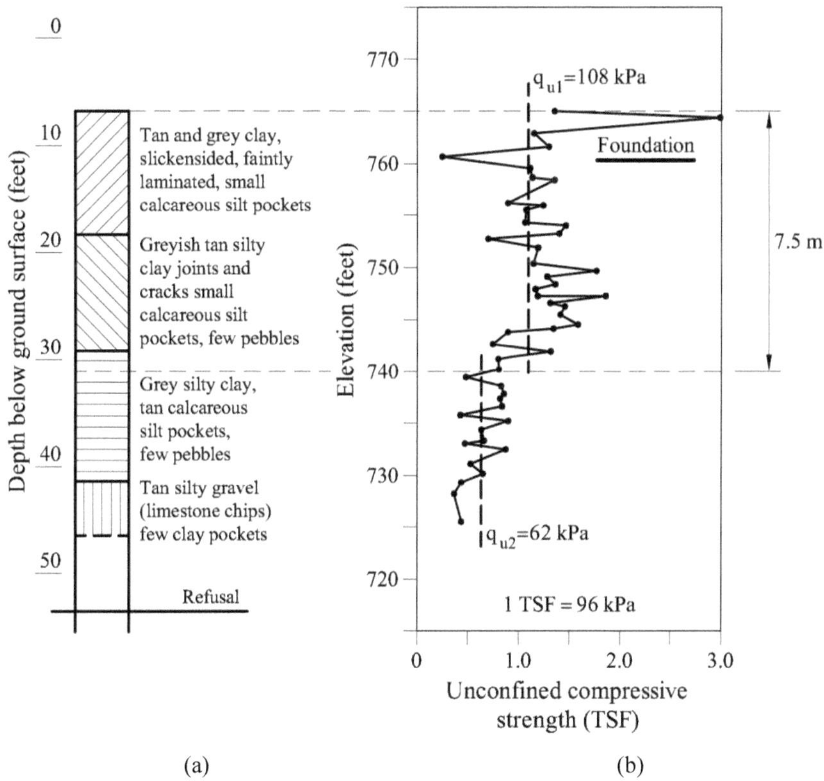

Figure 4.3 Soil profile below the elevator (after Peck and Bryant, 1953: © 1953 Thomas Telford Limited. Used with permission): (a) classification; (b) unconfined compression strength.

In terms of its colour, grain distribution and mineralogical content the clay may have looked almost homogeneous with depth (Fig. 4.3a). Note that these were the only properties that could be assessed in wash-borings in 1913. In contrast, in terms of the unconfined compression strength, q_u, there are two easily distinguishable layers (Fig. 4.3b). The upper one, a 7.5 m thick stiff clay layer with q_u = 108 kPa (undrained shear strength $c_u = q_u/2 = 54$ kPa), appears to be resting on a softer clay layer with q_u = 62 kPa ($c_u = 31$ kPa). This finding suggests that the elevator failure was most likely caused by the insufficient bearing capacity of its foundation.

4.1.4 The bearing capacity failure

The first three chapters of this book demonstrate how a structure can fail to function properly due to excessive settlements and tilt, which are caused by high compressibility of soil, rather than by its insufficient strength. In this chapter, we are concerned with the classical bearing capacity failure due to insufficient undrained shear strength. How could we distinguish between the two types of failure?

An excessive settlement takes place due to consolidation and secondary compression and is caused by a decrease of the soil volume below the foundation. Therefore, it takes place rather slowly and the ground surface around the structure subsides together with the foundation (Fig. 4.4a).

In a bearing capacity failure, a failure mechanism is formed below the foundation (Fig. 4.4b). The settlement takes place much faster and without decrease of the soil volume. Therefore, the displaced soil has to find itself an exit, causing a ground heave in the vicinity of the structure. This ground heave is a distinctive feature of the failure of the Transcona Grain Elevator (Fig. 4.4c), and was already present at the early failures of the railroad embankments described in Section 4.1.2.

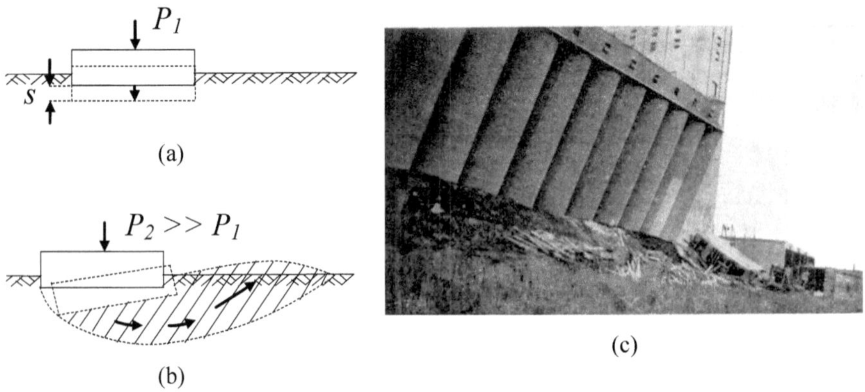

Figure 4.4 Bearing capacity failure: (a) settlement; (b) failure; (c) the ground heave (after White, 1953: © 1953 Thomas Telford Limited. Used with permission).

The particular problem of the Transcona Grain Elevator was that the failure mechanisms of the plate loading tests were apparently confined to the upper stiffer clay layer, due to the relatively small size of the plates. The elevator foundation, however, developed a much deeper failure mechanism which entered the weaker clay layer, significantly reducing the bearing capacity.

The problem of the bearing capacity of layered strata has received a lot of attention in the literature, using both rigorous analytical and numerical approaches (Merifield *et al.*, 1999). The purpose of this chapter is to show how some simple approximate methods provided a fairly good prediction for the bearing capacity failure of the Transcona Grain Elevator.

4.2 The Theory

4.2.1 Undrained bearing capacity formula

Under rapid loading, the ultimate bearing capacity of a $b \times L$ rectangular footing, based at a depth t on saturated clay with the undrained shear strength c_u, and the total unit weight γ, is given by the following modification of Terzaghi formula (Lang *et al.*, 2007):

$$\sigma_f = (\gamma \cdot t + q) + c_u N_c \left(1 + s_c + d_c\right), \tag{4.1}$$

where q is a surcharge load;

$$N_c = 2 + \pi \approx 5.14 \tag{4.2}$$

is the bearing capacity factor;

$$s_c = 0.2\frac{b}{L} \tag{4.3}$$

is the shape correction factor;

$$d_c = 0.4\frac{t}{b} \tag{4.4}$$

is the depth correction factor.

The safety factor against the bearing capacity failure for a footing with a contact pressure σ is thus given by

$$F_s = \frac{\sigma_f}{\sigma}. \tag{4.5}$$

4.2.2 Upper bound limit analysis

The value of the bearing capacity factor in Equation (4.2) can be obtained using the upper bound limit analysis of the Prandtl (1920) kinematic failure mechanism for a strip footing (Fig. 4.5a). This mechanism consists of two rigid triangular blocks and a fan shear zone.

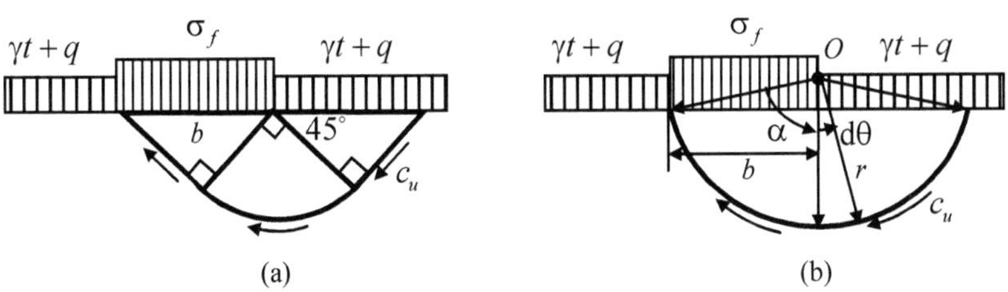

Figure 4.5 Kinematic failure mechanisms: (a) Prandtl; (b) simple scoop.

For a bearing capacity problem, the Upper Bound Theorem of the Limit Analysis (Drucker *et al*., 1952) states that *collapse will occur under the smallest values of the surface loads for which it is possible to find a kinematically admissible failure mechanism.* For plastic deformation governed by Tresca's yield criterion (applicable to undrained loading conditions), a failure mechanism is said to be kinematically admissible if

- it satisfies the incompressibility condition;
- it satisfies any imposed kinematic boundary conditions.

It can be shown that the Prandtl mechanism (Fig. 4.5a) is kinematically admissible.

In simple words, if we find a kinematically admissible mechanism and determine the surface load which brings it to failure, this load will be always larger than the true collapse load (or in the best case equal to it). To find these surface loads we use the fact that, in the absence of inertial effects, the rate at which the applied loads do work (calculated by multiplying these loads by the rates of displacements at the points of load application) must equal the rate of internal plastic work. We always try to find a mechanism which fails at the smallest surface load and, if we are lucky, this smallest load will be equal to the largest load obtained from the Lower Bound Theorem of the Limit Analysis (Drucker *et al*., 1952). In this case it will represent the true collapse load, as is the case with the Prandtl mechanism in Figure 4.5a, so that $N_c \approx 5.14$ in Equation (4.2) is actually the exact solution.

To demonstrate how the Upper Bound Limit Analysis works, we shall, however, consider a much simpler, rigid scoop mechanism in Figure 4.5b, with radius r and the center of rotation O located over the footing edge. It satisfies kinematic boundary conditions (nothing obstructs its rotation). Being rigid, it also automatically satisfies the incompressibility condition. Therefore it is kinematically admissible. Let us find its failure surface load σ_f.

Assume that the mechanism rotates counter-clockwise by a small angle $d\theta$ (Fig. 4.5b). The work of the surface loads on this small rotation can be then calculated as a product of forces and the corresponding displacements:

$$W = \left[\sigma_f b - (\gamma t + q) b\right]\left(\frac{b}{2} d\theta\right). \tag{4.6}$$

The internal plastic work dissipated on the scoop boundary during this small rotation is calculated as a product of the shear resistance integrated along the circular arc and the displacement on the boundary

$$W^P = (2\alpha r c_u)(r d\theta). \tag{4.7}$$

Equating (4.6) and (4.7), we can express the failure load σ_f in the form of the equation

$$\sigma_f = (\gamma \cdot t + q) + c_u N_c, \tag{4.8}$$

where

$$N_c = 4\alpha\left(r/b\right)^2. \tag{4.9}$$

This failure load is larger than the true collapse load. However, by looking for the most dangerous kinematic mechanism from the rigid scoop family in Figure 4.5b, we can minimize this failure load and approach the true collapse load from above. Noting that $\sin\alpha = b/r = x$, we can rewrite (4.9) as

$$N_c = 4x^{-2}\arcsin x \tag{4.10}$$

and the smallest failure load will be obtained for *x*, at which $\partial N_c/\partial x = 0$:

$$-8x^{-3}\arcsin x + 4x^{-2}\big/\sqrt{1-x^2} = 0. \tag{4.11}$$

Equation (4.11) has a solution at $x \approx 0.919$, which corresponds to the most dangerous mechanism in this family, with the center of rotation O located at the distance $b\sqrt{1/x^2-1} \approx 0.43b$ above the footing edge (Fig. 4.5b). It corresponds to the smallest failure load of $N_c \approx 5.52$. As expected, this load is larger than the true collapse load of $N_c \approx 5.14$, but only by 7%, despite the fact that the scoop mechanism is considerably simpler than the Prandtl mechanism! The real advantage of the scoop mechanism will, however, become evident when we consider two-layer strata.

4.2.3 Two-layer strata

For our bearing capacity calculations, we have only so far considered homogeneous soils. Consider now two-layer strata (Fig. 4.6), with an upper layer of thickness D and undrained shear strength c_{u1}, and the lower layer of thickness larger than $b-D$ and undrained shear strength c_{u2}.

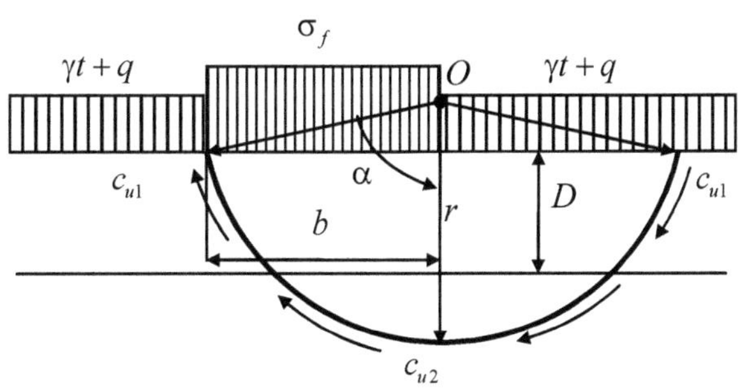

Figure 4.6 A two-layer strata: a kinematic failure mechanism.

For the Prandtl mechanism (Fig. 4.5a), it becomes very difficult to calculate the failure load, and it is even not guaranteed that this is going to be the exact solution.

An approximate solution can be found by weighted averaging of the undrained shear strength over the depth $b/2$ (approximate depth of the failure mechanism):

$$c_u = c_{u1}m + c_{u2}(1-m) = c_{u1}(m+n-mn), \tag{4.12}$$

where

$$m = 2D/b, \qquad n = c_{u2}/c_{u1} \tag{4.13}$$

and substituting it into Equation (4.1). Though often used in practice, this is not a rigorous plasticity theory solution, and we do not even know if the obtained bearing capacity is below or above the true failure load.

The scoop mechanism (Fig. 4.6), however, does offer a rigorous upper bound. The work of the surface loads on a small rotation $d\theta$ is still given by Equation (4.6). The internal plastic work dissipated during this small rotation on the scoop boundary now has two terms:

$$W^P = \left(2\alpha r c_{u1} + 2\arccos\left(\frac{\sqrt{r^2-b^2}+D}{r}\right) r\left(c_{u2}-c_{u1}\right)\right)\left(r d\theta\right). \tag{4.14}$$

Equating (4.6) and (4.14) we can express the failure load σ_f in the form of the equation

$$\sigma_f = (\gamma \cdot t + q) + c_{u1}N_c, \tag{4.15}$$

where

$$N_c = 4\left(\alpha + (n-1)\arccos\left(\sqrt{1-\left(\frac{b}{r}\right)^2} + \frac{m}{2}\frac{b}{r}\right)\right)\left(\frac{r}{b}\right)^2. \tag{4.16}$$

Again, by looking for the most dangerous kinematic mechanism, we shall try to minimize this failure load and approach the true collapse load from above. Because $\sin\alpha = b/r = x$, we can rewrite Equation (4.16) as

$$N_c = 4x^{-2}\left(\arcsin x + (n-1)\arccos\left(\sqrt{1-x^2} + \frac{mx}{2}\right)\right) \tag{4.17}$$

and the smallest failure load will be obtained for *x*, at which $\partial N_c/\partial x = 0$:

$$-8x^{-3}\left(\arcsin x+(n-1)\arccos\left(\sqrt{1-x^2}+\frac{mx}{2}\right)\right)+$$
$$4x^{-2}\left(\frac{1}{\sqrt{1-x^2}}-\frac{(n-1)\left(-\frac{x}{\sqrt{1-x^2}}+\frac{m}{2}\right)}{\sqrt{1-\left(\sqrt{1-x^2}+\frac{mx}{2}\right)^2}}\right)=0. \tag{4.18}$$

Equation (4.18) can be resolved numerically with respect to x, which would correspond to the most dangerous mechanism for specific values of m and n. Substitution of this x into Equation (4.17) will produce the smallest failure load. For a weaker lower layer $c_{u2} \le c_{u1}$ $(0 \le n \le 1)$ and $0 \le m \le 3$, the corresponding values of N_c are given in a plot in Figure 4.7 (after Button, 1953).

The bearing capacity factor cannot exceed $N_c \approx 5.52$ corresponding to homogeneous soil.

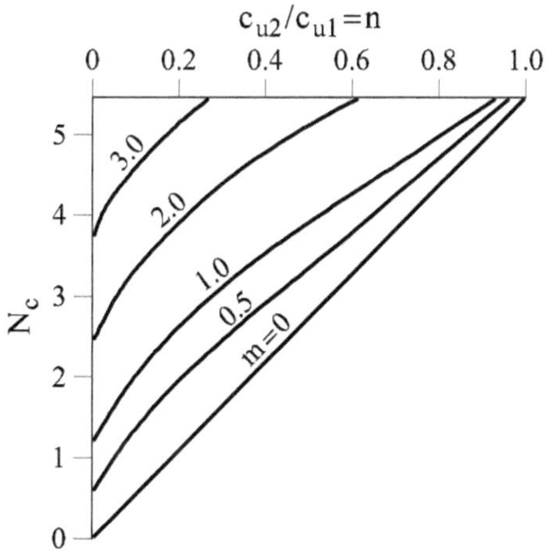

Figure 4.7 A two-layer strata: the bearing capacity factor for $c_{u2} \le c_{u1}$ (modified from Button, 1953).

4.2.4 Summary

For homogeneous soils, the conventional bearing capacity formula based on the Prandtl failure mechanism provides an exact solution (of course, within the assumptions of the theory of plasticity – nothing in geomechanics is truly "exact"). For two-layer strata, however, this solution cannot be applied directly, and its approximation using a weighted average of shear strength with depth is not rigorous. This is where the Upper Bound Limit Analysis can provide some useful results, using simpler kinematic failure mechanisms, such as the scoop mechanism in Figure 4.6. The values of the bearing capacity factor in the plot in Figure 4.7, are rigorous upper bounds and are, of course, larger than the true collapse load. Therefore, they are not conservative and an important question has to be answered: how far are we from the true collapse load? Analysis of the Transcona Grain Elevator failure provides us with a unique opportunity to make this assessment basing of a full-scale case study.

4.3 The Analysis

The theory presented above provides tools for a simplified bearing capacity analysis of the Transcona Grain Elevator foundation, which follows the approach of Perloff and Baron (1976).

4.3.1 Model parameters

Geometry (Fig. 4.2a and Peck and Bryant, 1953):

- width of the rectangular footing: $b = 23.5$ m;
- length of the rectangular footing: $L = 59.5$ m;
- depth of the foundation: $t = 3.7$ m.

Soil profile (Fig. 4.3 and Peck and Bryant, 1953):

- thickness of the upper clay layer below the footing $D = 6.0$ m;
- undrained shear strength of the upper layer $c_{u1} = q_{u1}/2 = 54$ kPa;
- undrained shear strength of the lower layer $c_{u2} = q_{u2}/2 = 31$ kPa;
- total unit weight of clay in both layers $\gamma = 18.7$ kN/m^3.

Loads (Allaire, 1916; Peck and Bryant, 1953):

- surface surcharge load $q = 0$;
- failure contact pressure from the plate load tests: $\sigma_f \approx 400$ kPa;
- true contact pressure at failure: $\sigma_f = 293$ kPa.

4.3.2 The bearing capacity assumed in the original design

In the original design, it was assumed that the soil profile was homogeneous with the properties of the stiff upper layer $c_{u1} = 54$ kPa. In this case, the bearing capacity of the foundation could be calculated using Equation (4.1), where the bearing capacity factor, the shape and depth correction factors are given by Equations (4.2) – (4.4):

$$N_c \approx 5.14, \quad s_c = 0.2\frac{23.5}{59.5} = 0.08, \quad d_c = 0.4\frac{3.7}{23.5} = 0.06 \tag{4.19}$$

so that

$$\sigma_f = (18.7 \times 3.7 + 0) + 54 \times 5.14 \times (1 + 0.08 + 0.06) = 386 \text{ kPa}. \tag{4.20}$$

This bearing capacity estimate is rather close to the smallest failure contact pressure of $\sigma_f \approx 400$ kPa obtained from the plate load tests, whose shallow kinematic mechanisms apparently stayed entirely within the upper layer. It is, however, significantly higher than the true failure pressure $\sigma_f = 293$ kPa.

4.3.3 A conservative estimate

Note that if the soil was homogeneous but with the properties of the weaker lower layer ($c_{u2} = 31$ kPa), the resulting bearing capacity would be

$$\sigma_f = 69.2 + 31 \times 5.14 \times 1.14 = 251 \text{ kPa}. \tag{4.21}$$

If this value was available to designers, the result would actually not be that bad: not only the elevator would not fail, it would not even be too much overdesigned. Indeed, the predicted value is only 20% more conservative than the true failure pressure of $\sigma_f = 293$ kPa.

4.3.4 Two-layer strata

A more-sophisticated analysis, based on a two-layer model, should produce more accurate predictions. First, we follow Peck and Bryant (1953) and use the approximate method based on the Prandtl solution using a weighted average of the undrained shear strength. To find its value, the parameters

$$m = 2 \times 6.0/23.5 = 0.51, \qquad n = 31/54 = 0.57, \tag{4.22}$$

are calculated from Equation (4.13) and substituted into (4.12):

$$c_u = 54 \times (0.51 + 0.57 - 0.51 \times 0.57) = 43 \text{ kPa}. \tag{4.23}$$

Then the bearing capacity is estimated as

$$\sigma_f = 69.2 + 43 \times 5.14 \times 1.14 = 321 \text{ kPa}, \tag{4.24}$$

which is 10% larger than the true failure pressure of $\sigma_f = 293$ kPa!

While the Prandtl solution for a homogeneous soil is the exact solution, in our two-layer approximation, we observe that it is not only inaccurate, but it is also not conservative and could lead to failure.

The scoop mechanism, in contrast, provides a remarkably good bearing capacity estimate. Using parameters (4.22), from the plot in Figure 4.7 (Button, 1953), we obtain

$$N_c \approx 3.7, \tag{4.25}$$

which, being substituted into (4.15), gives

$$\sigma_f = 69.2 + 54 \times 3.7 \times 1.14 = 297 \text{ kPa}. \tag{4.26}$$

Being an upper bound, this value, as expected, is higher than the true failure pressure of $\sigma_f = 293$ kPa, but only marginally. It would provide an excellent estimate for design of the elevator, if only the Soil Mechanics was more mature in those days and the soil properties were properly determined.

4.3.5 Discussion

The above analysis confirms that insufficient bearing capacity was the most likely cause of the Transcona Grain Elevator failure. The Prandtl mechanism would provide a reasonably good prediction of the bearing capacity if the soil was homogeneous, as confirmed by the plate load tests, where the failure mechanism was entirely confined to the upper clay layer. The real mechanism was much deeper due to the large foundation width and penetrated a weaker lower layer. An approximate approach using the Prandtl formula with averaged shear strength appeared to be neither accurate nor conservative. In contrast, the upper bound limit analysis using a scoop mechanism provided a remarkably good prediction.

4.4 Mitigation Measures

4.4.1 Emptying of the elevator

The first priority was to save the wheat. This was done by tapping holes in the most westerly row of bins at approximately ground level and bleeding out the grain upon a belt conveyor built parallel to the bins (Fig. 4.8). After the bins of this outer row were emptied down to the ground line, holes were made in the next row, and so on. This was a dangerous operation, because the bins were not designed for such an inclination and for the varying load of grain successively emptied row after row. The grain below the ground level was removed to the deep trench dug at the north side using existing underground conveyor tunnels. In spite of the difficulty and danger of these operations, all wheat was successfully saved in less than 3 weeks.

Figure 4.8 Emptying the grain (after White, 1953: © 1953 Thomas Telford Limited. Used with permission).

4.4.2 Underpinning of the work-house

The next step was underpinning the work-house to rock, as it was feared that it may become unstable as well. This was achieved by sinking a 1.5 m diameter pier under each of the 24 columns of the building. Because of the heavy loads, the height of the structure and its small base, it was necessary to reinforce the foundation and the structure before the underpinning operation could begin. This was achieved by sinking twenty 1.2 m diameter piers outside the building walls and constructing a framework of wooden pushers. A truss arrangement was formed allowing for a larger part of the building load to be temporarily transferred to the external piers. After this, access tunnels were excavated under the foundation mat and 2 m diameter piers were sunk down to rock under each of the internal columns. The 1.5 m diameter piers were sunk under the wall columns and the building load was transferred to the new foundation. The work was completed by June 1914.

4.4.3 Straightening of the bin-house

The works to straighten and underpin the bin-house started in February 1914. It was decided not to return to its original elevation, but just to bring it above the groundwater level, 4.3 m below its original position. A row of fourteen 2.1 m diameter piers below the lower end of the foundation slab was sunk to the rock and concreted (Fig. 4.9a). As the under-excavation on the east side commenced (Fig. 4.9b), the building began to rotate around the piers, assisted by the row of pushers on the west side.

Four additional rows, each of fourteen 2.1 m diameter piers, were sunk to the rock level (Fig. 4.10a). As the building continued to get straightened by excavating on the higher side, the center of rotation was first shifted to the second pier row on the lower side, and then to the central row. The jacking screws (Fig. 4.10b) were then placed at the freed western pillars and helped to control further rotation by pushing the western side of the building up.

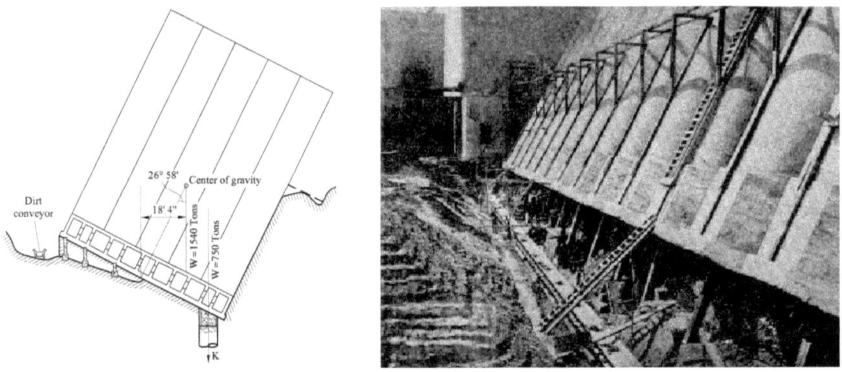

(a) (b)

Figure 4.9 Under-excavation at the east (high) side (after Allaire, 1916): (a) a cross-section with the piers; (b) a view.

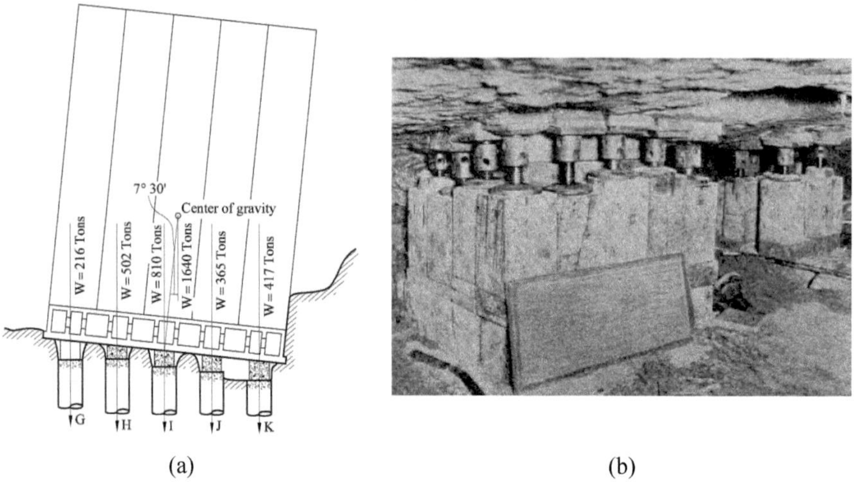

Figure 4.10 Jacking-up the west (low) side (after Allaire, 1916): (a) a cross-section with the piers; (b) the screw jacking units.

4.4.4 Underpinning of the bin-house

In October 1914, the building returned to the vertical position (Fig. 4.11). All five rows of 14 deep foundation shafts each were already in place, and all that was left to do was to remove the jacking units and to concrete the piers up to the contact with the original foundation slab. It was decided not to straighten a slight 1 degree tilt of the bin-house towards the north (Fig. 4.11), as it did not interfere with the elevator operation.

Figure 4.11 The elevator in righted position (after White, 1953: © 1953 Thomas Telford Limited. Used with permission).

4.5 Lessons Learned

4.5.1 Site investigation

The main lesson is the importance of the proper geotechnical site investigation. The fact that other similar heavy structures in the area were founded on similar foundations did not help. A locally weaker clay layer located below the typical stiff clay layer messed up all the analogies.

4.5.2 Field load tests

The field plate load tests also did not help to avoid the failure. This happened due to the scale effect: the small size of the plates compared to that of the foundation did not allow for the deeper weak clay layer to be involved in the failure mechanism during the tests. Conclusion: in order to be able to make meaningful predictions based on the field load tests, these tests have to be performed at the real one-to-one scale or at different elevations (which is an involved and expensive operation).

4.5.3 Conservative design

The Terzaghi formula based on the Prandtl mechanism provides the exact solution for undrained bearing capacity of a homogeneous soil. For two-layer strata, application of this formula with averaged shear strength may not necessarily be conservative. In this case a conservative prediction may be achieved by assuming that the soil is homogeneous with the undrained shear strength of the weaker layer. This prediction, however, may lead to an over-conservative and too expensive design.

4.5.4 Upper bound limit analysis

An upper bound limit analysis using a simple kinematic mechanism can provide a useful tool for determining an upper bound for bearing capacity of a foundation on layered strata. Though the obtained prediction is not conservative, in some cases (like the one considered in the present chapter), it may be pretty close to the true failure load.

References

Allaire, A. (1916) The failure and righting of a million-bushel grain elevator. *Proceedings of ASCE* XLI, 10, 2535 – 2568.

Button, S.J. (1953) The bearing capacity of footings on a two-layer cohesive subsoil. *Proceedings of the 3rd International Conference on Soil Mechanics and Foundation Engineering*, Zürich, 1, 332 – 335.

Drucker, D.C., Prager, W. and Greenberg, H.J. (1952) Extended limit design theorems for continuous media. *Quarterly of Applied Mathematics* 9, 381 – 389.

Engineering News (1913) Failure of Transcona Grain-Elevator. *Engineering News* 70 (19), 944 – 945.

Lang, H.J., Huder, J., Amann, P. and Purzin, A.M. (2007) *Bodenmechanik und Grundbau*. Springer-Verlag, Berlin, 353 pp.

Merifield, R.S., Sloan, S.W. and Yu, H.S. (1999) Rigorous plasticity solution for the bearing capacity of two-layered clays. *Géotechnique* 49 (4), 71 – 490.

Peck R.B. and Bryant F.G. (1953). The bearing capacity failure of the Transcona Elevator. *Géotechnique* 3 (5), 201–208.

Perloff, W.H. and Baron, W. (1976) *Soil Mechanics. Principles and Applications*. John Wiley & Sons, New York, 745 pp.

Prandtl, L. (1920) Über die Härte plastischer Körper. *Nachrichten von der Gesellschaft der Wissenschaften zu Goettingen, Mathematisch-Physikalische Klasse*, 37: 74–85.

White, L.S. (1953) Transcona elevator failure: Eye witness account. *Géotechnique* 3 (5), 209–214.

Chapter 5

Caisson Failure Induced by Liquefaction: Barcelona Harbour, Spain

TABLE OF CONTENTS

A.M. Puzrin et al., *Geomechanics of Failures*, DOI 10.1007/978-90-481-3531-8_5,

Chapter 5

Caisson Failure Induced by Liquefaction: Barcelona Harbour, Spain

5.1 Building a Caisson Dyke

The design of a new entrance for the Barcelona harbour involved opening a channel through the existing dyke and the protection of the opening by means of a new dyke made of reinforced concrete caissons (Fig. 5.1).

Caissons (each 19.6 m wide, 19.5 m high and 33.75 m long) were built in a mobile platform and towed to their intended position, shown in Figure 5.1. Caissons have a cellular structure. Inner vertical concrete walls allow filling the caisson in a controlled manner. In this way, the caisson may be precisely sunk (by controlled inundation of cells). Once "in situ", the total weight is increased by sand filling the caisson cells. Caisson foundation design should ensure stability against caisson weight and wave loading.

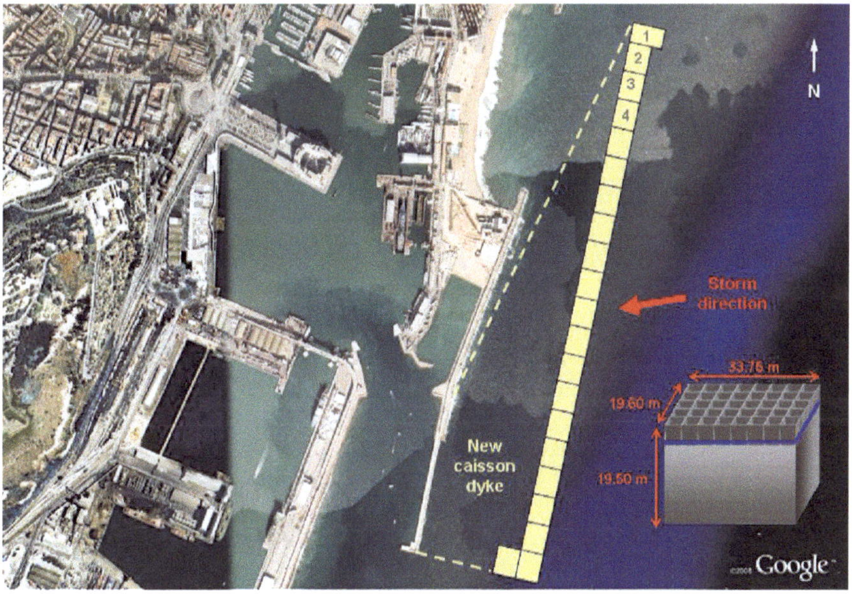

Figure 5.1 The new caisson breakwater (background photograph from Google Earth).

Foundation soils were deposited during the development of two overlapping deltas (Besós river delta towards the north-east and Llobregat river delta towards the south-west of the site). Soft silts and silty clays extend from the surface to substantial depths (tens of meters). A band close to the coastline is covered by a mantle of sand whose thickness decreases towards the sea.

The deep soft soils in the Barcelona harbour area are a challenge for caisson stability. The favoured design is to substitute part of the natural soils by a frictional fill extending on both sides of the breakwater (sea side and land side). Figure 5.2 shows a sketch of the foundation conditioning. A dredged trench is first excavated. Coarse granular soils are then backfilled and a final gravelly layer is leveled in preparation for caisson sinking.

Once sunk in place, caissons are finally capped with a concrete slab and a protective wall is built to avoid wave overrunning. It will be shown later that the vertical average net stress of the caissons filled with sand against the foundation soils is of the order of 220 kPa.

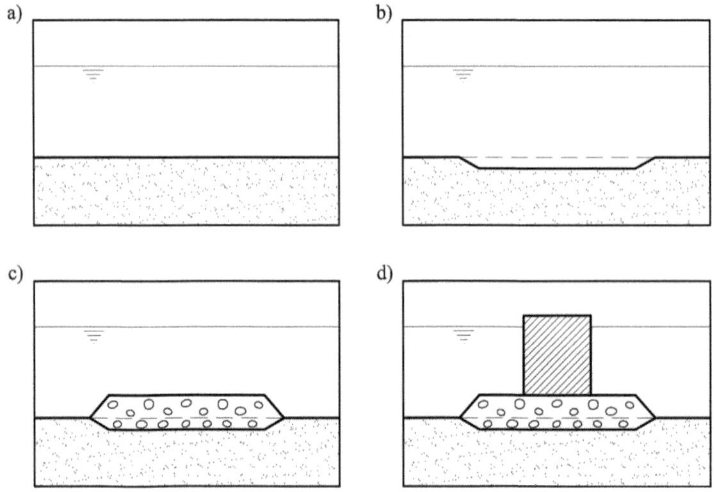

Figure 5.2 Caisson construction sequence: (a) initial soil conditions; (b) trench excavation; (c) extension of granular embankment; (d) caisson sinking.

5.2 The Failure

Dredging of the trench prepared to receive the coarse granular fill was finished on November 2000. Trench filling took the following six months. On May 10, 2001 the granular base was levelled and ready for the sinking of four caissons, in the position shown in Figure 5.1 (Caissons 1, 2, 3 and 4). Caisson sinking began, however, in the middle of October 2001. Cells were filled with sand some days later.

On November 10, 2001 an east-northeast storm with maximum significant wave heights of 4 m hit the coast. The time record of wave period and significant height is given in Figure 5.3. Some time during the night of November 10 and November 11, the four caissons failed. Figure 5.4 shows an aerial view of the failure. The two central caissons are not in sight and the extreme ones are seen to be tilted and partially submerged.

This failure was not a good starting experience for a breakwater typology which began to be used in the Barcelona harbour area, known for its soft foundation soils. The traditional and successful design was the embankment type of breakwater. The failure teaches, however, an important lesson to geotechnical engineers: the risk involved in moving ahead of standard well-proven engineering practices and entering into new ground, into a "terra incognita".

Let us examine first the failure in more detail.

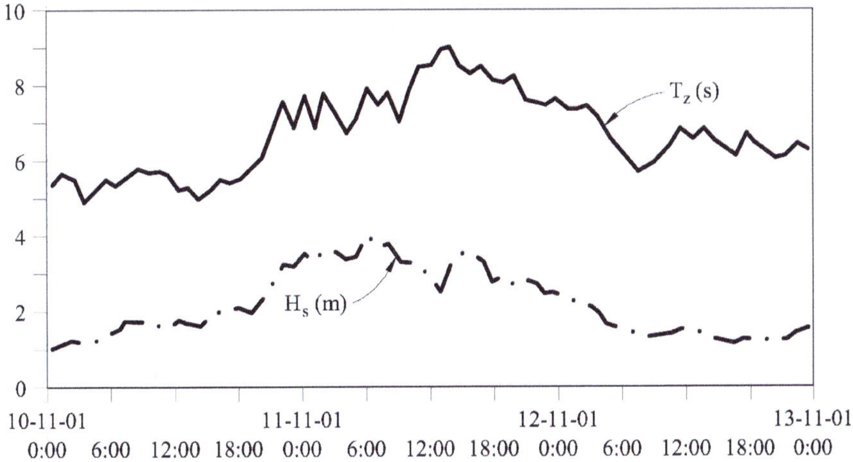

Figure 5.3 Significant wave heights (H_s) and wave period (T_z) of the storm.

Figure 5.4 Failed caissons.

The recorded maximum wave period was 9 seconds. The maximum intensity of the storm in terms of significant wave height (H_s = 4 m) lasted around one hour and therefore the number of wave load applications during this time interval was around 250 – 350. However, the precise failure time is unknown.

Soil profiles were established after the failure. They could be compared with the sea bottom topography before the works and immediately before caisson installation. Such a comparison is given in Figure 5.5 for a cross-section of Caisson 3 (one of the central caissons). The original and final positions of the caisson are also plotted.

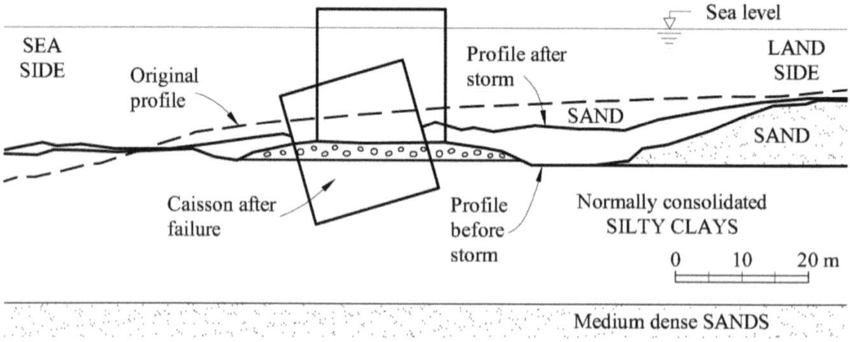

Figure 5.5 Cross-section through Caisson 3 before and after failure. Also shown is the original soil profile, the excavated profile, the granular berm, and the soil profile after the failure.

The caisson is deeply buried into the soil. The tilt of caisson top, towards the open sea, is consistent with a bearing capacity type of instability induced by an inclined load (the resultant of caisson self-weight and wave loading).

The caisson volume below the soil surface is estimated in Figure 5.5 to be 240 m^3/m. The depth of burial suggests that the foundation soil could have liquefied. This aspect will be examined later. The internal caisson walls were severely damaged. Wall reinforcement was not intended to resist the efforts associated with a large tilt.

The four caissons involved in the failure were later covered by a conventional fill-type breakwater. However, the remaining caissons envisaged in the project were built after a revision of the foundation design. They provided settlement data, shown later, which helps to derive some foundation parameters (average stiffness and consolidation coefficient).

The profiles given in Figure 5.5 indicate that the initial excavation in sands, in the land direction, was substantially filled again after the caisson failure. The calculated soil volume between the surface profiles before and after the storm is around 220 m^3/m, a value which is very similar to the buried caisson volume under the foundation level. It is then reasonable to accept that the caisson failure displaced the foundation soil towards the land side following a deep failure surface. It is also inferred that wave action after the caisson failure distributed the volume of soil initially displaced by the caisson failure over a wider area.

5.3 Soil Conditions

Figure 5.6 shows a simplified representative stratigraphic profile under caissons. An upper layer of loose silty sand, 10 meters thick, overlies a deposit of clayey silts and silty clays, 20 m thick. Below, a level of medium to dense sands was found. The upper 9 m of sands were removed by dredging. It appears that the thickness of the coarse granular fill below the caisson was rather small (around 2 m). The figure also shows the estimated lateral extent of the coarse granular embankment in the land direction. A detailed stratigraphic record with additional information on soil parameters derived from a few undisturbed samples tested is given in Figure 5.7. The figure also includes SPT N values.

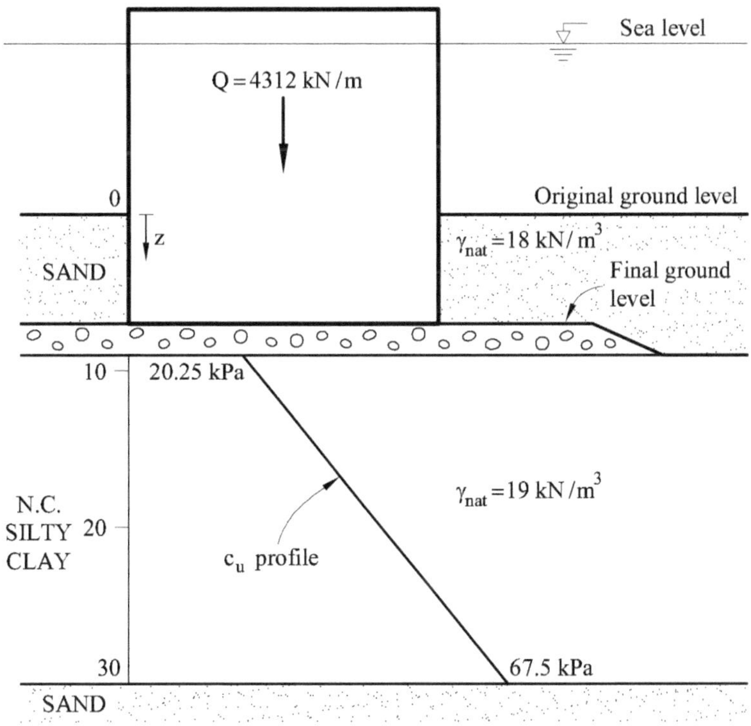

Figure 5.6 Simplified soil profile under caissons.

The silty clay is a soft deposit as revealed by the low N values (9, 4, 4, 5, 13…). It has a moderate plasticity (w_L= 30 – 32.6%) and the Plasticity Index is particularly low (4 – 10%). These deltaic deposits classify as ML, CL-ML or CL. The void ratio is high: 0.92 – 0.96.

Figure 5.8 shows an oedometric compression curve of a specimen recovered at a depth of 12.50 m below the soil surface. If normally consolidated, the vertical yield or preconsolidation stress would be around 12.50 m $\times$ 8 kN/m^3 = 100 kPa. This is close to the value found in the oedometer test using a classical construction shown in the figure. It is concluded that the silt deposit is normally consolidated.

The calculated virgin compression coefficients (C_c= 0.22 – 0.26) are high.

The coefficient of consolidation determined in small specimens is of limited reliability. The settlement records of the caissons built later will be analyzed below to determine this parameter and to estimate the permeability "in situ". The lower silty sands and clean sands are markedly stiffer. This is reflected in the high SPT values (N = 15 to 46), in the lower void ratio (e = 0.7) and in the small compressibility index, C_c = 0.06, measured in an oedometer test on a recovered sample.

Unconfined compression strengths, measured in samples (12 – 19 kPa), remain below the minimum accepted values for a normally consolidated low plasticity deposit, a result which may be explained by sample disturbance.

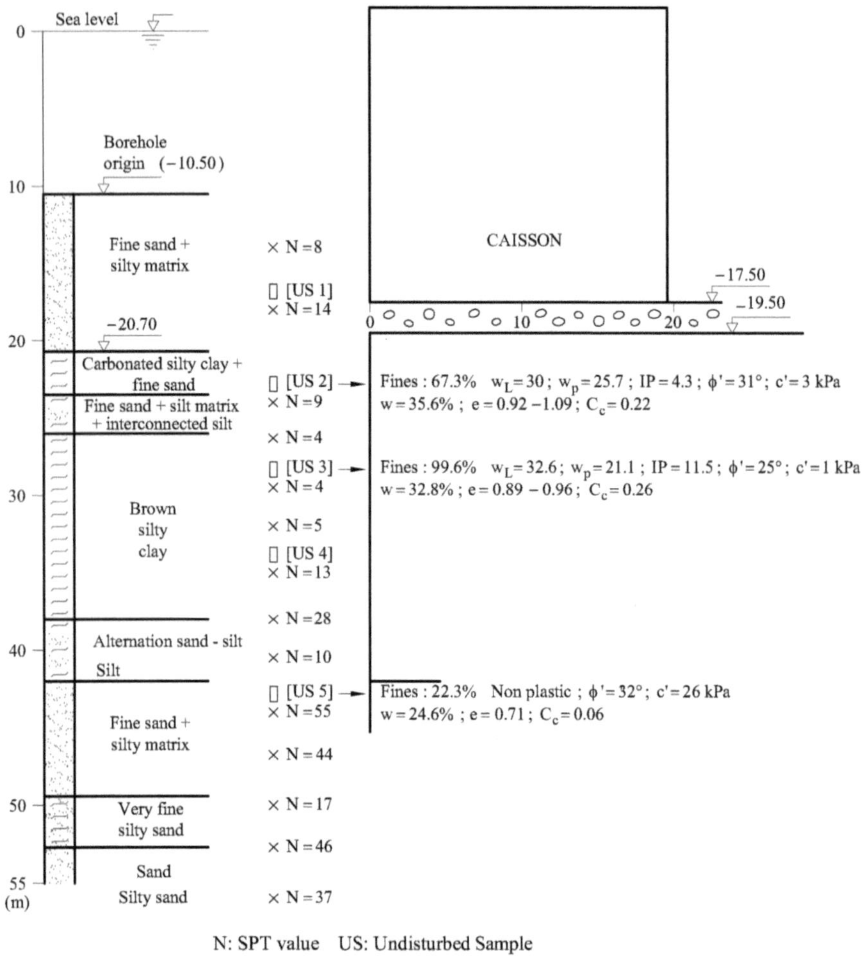

Figure 5.7 Detailed soil profile under caissons.

The undrained strength of normally consolidated soils increases with the confining effective stress. A useful relationship is

$$c_u = a\sigma'_v, \tag{5.1}$$

where σ'_v is the vertical effective stress and a is a coefficient which takes values in the range $a = 0.25 - 0.30$. Expressions have also been found for c_u in terms of the mean effective stress, σ'_m, which are also useful in applications

$$c_u = \bar{a}\sigma'_m. \tag{5.2}$$

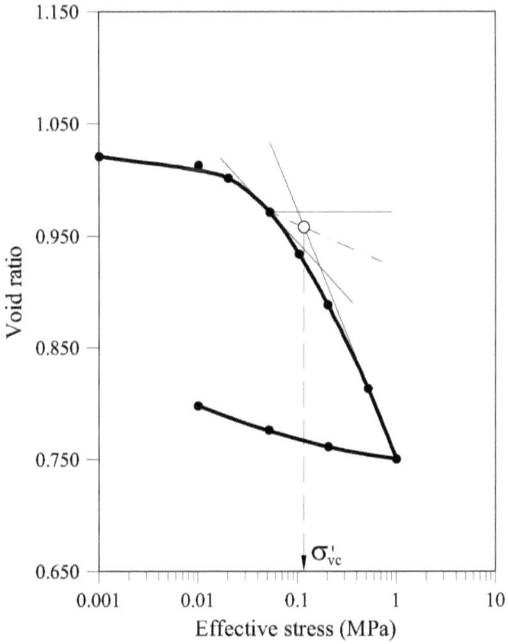

Figure 5.8 Oedometer test on a sample recovered at a depth of 12.50 m below the soil surface.

The expression for coefficient a and $\bar{a}$ can be derived following a theoretical procedure. For instance, Wood (1990) and Potts and Zdravkovic (1999) presented such derivations for a Cam Clay elastoplastic model (see Chapter 6). If the coefficient of earth pressure at rest, K_0, is known, the mean effective stress is given by

$$\sigma'_m = \frac{1+2K_0}{3}\sigma'_v \tag{5.3}$$

and, therefore,

$$\bar{a} = \frac{3a}{1+2K_0}. \tag{5.4}$$

For a normally consolidated clay, K_0 = 0.5, and $\bar{a} = 0.38$ if $a = 0.25$. This value for $\bar{a}$ will be used later. Further discussion on undrained strength is given in Chapter 6.

Simple shear tests performed in specimens recovered in the same deltaic formation of silty soils provided c_u/σ'_v values in the range 0.25 – 0.30. It appears that Equations (5.1) or (5.2) provide a good estimation of undrained strength in our case. It indicates that strength increases linearly with depth. The excavation of upper sands leaves the soil overconsolidated (provided that enough time has elapsed to dissipate pore pressures) but the undrained strength remains slightly below its original value because water content changes during unloading will remain small and the soil void ratio will essentially remain unchanged.

It follows that the first sinking of the caisson found a silty soil on the upper boundary of the silty clay deposit (at 9 m of depth with respect to the original ground level) having a value of c_u of around

$$c_u \approx 0.25\sigma'_v = 0.25 \cdot \gamma_{\text{sub}} \cdot \text{depth} = 0.25 \times 9\ \text{kN/m}^3 \times 9\ \text{m} = 20.25\ \text{kPa}, \tag{5.5}$$

where γ_{sub} is the average submerged unit weight for sands and silts that correspond to an average saturated unit weight (γ_{sat}) of 19 kN/m^3 and a water unit weight of 10 kN/m^3. On the lower contact between the silty layer and the dense sand level on the bottom (at a depth of 30 m), the undrained strength is

$$c_u = 0.25 \times 9\ \text{kN/m}^3 \times 30\ \text{m} = 67.5\ \text{kPa}. \tag{5.6}$$

The distribution of c_u with depth is plotted in Figure 5.6.

The remaining properties indicated in Figure 5.7 complete the description of the soil. Drained direct shear tests provided friction angles of $25° - 31°$ and negligible cohesion intercepts.

Additional data was provided by a cone penetration test (CPT) performed in the caisson foundation area during the design stage. The test was run at a water depth of 24 m on the sea side of the breakwater position. The record is shown in Figure 5.9. The test was run in several stages from the bottom of an advancing borehole. The initial penetration resistances at every repeated pushing operation are affected by a stress release induced by the boring excavation and possibly by some soil remoulding. If these initial parts of the penetration records are disregarded, the test shows a linear increase of the cone penetration resistance with depth, which is an indication of a normally consolidated state of the soil. Being at a water depth of 24 m the cone is recording the strength of silty clays (the upper sand layer is not present at these water depths). The undrained strength is correlated with point resistance, q_c, through (Lunne *et al.*, 1997)

$$c_u = \frac{q_c - \sigma_v}{N_k}, \tag{5.7}$$

where σ_v is the total vertical stress at the current location and N_k is a "bearing capacity" factor, which takes values in the range 10 – 20. For the Barcelona harbour soils, a value $N_k = 15$ provides c_u values consistent with undrained simple shear data. The CPT test in Figure 5.9 is a good indication of the normally consolidated conditions of the silt layer and, also, on the validity of Equation (5.1) with $a = 0.25$. The peak resistance values indicated in Figure 5.9 correspond to more resistant and dilatant sand layers. The silty clay strength corresponds to the minimum envelope of the q_c record, leaving aside the peaks and the disturbed initial parts of successive records.

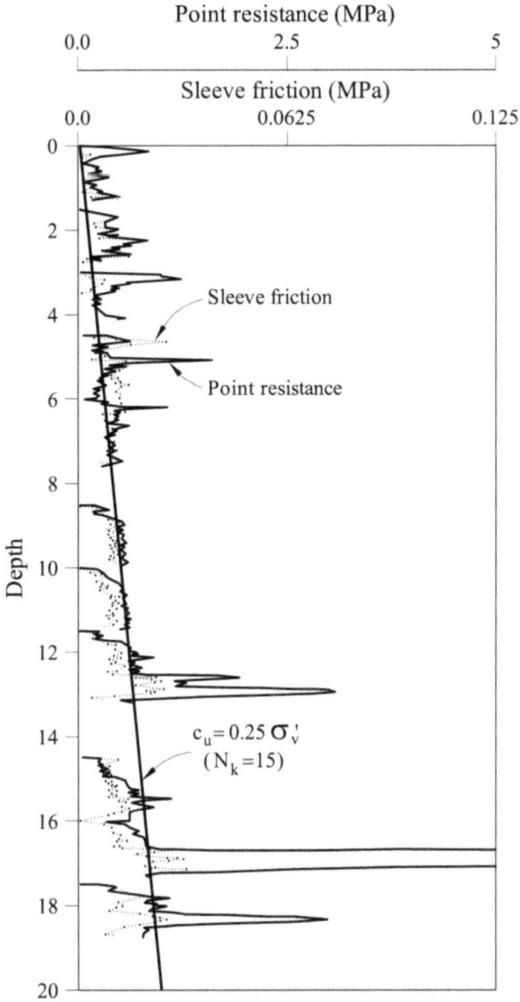

Figure 5.9 Cone penetration resistance and sleeve ratio of CPT test on foundation soils.

5.3.1 Liquefaction

The susceptibility to liquefaction of the low plasticity silty soil of the caisson foundation may be evaluated by performing undrained cyclic shear tests. But there is also a possibility of profiting the accumulated experience in earthquake engineering. A survey on the relationship between risk of liquefaction (under earthquake conditions) and type of soil (identified by its plasticity) is given in Figure 5.10 (Seed *et al.*, 2003). The position of samples represented in Figure 5.7 is also indicated in the plasticity chart. They fall in the area of "potentially liquefiable" materials.

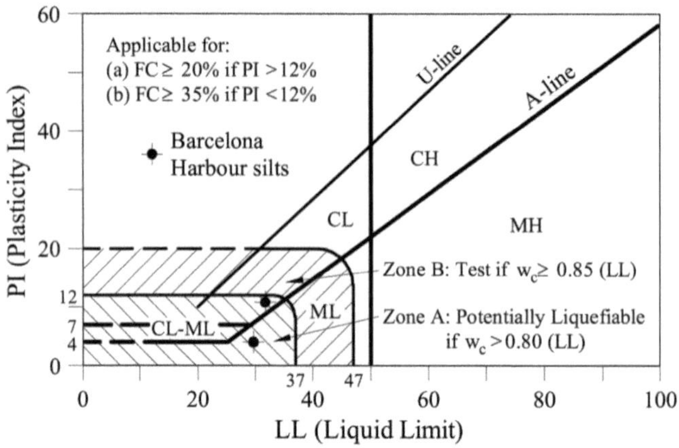

Figure 5.10 Criteria to assess the possibility of liquefaction of fine soils (Seed *et al.*, 2003. With permission from ASCE). Also indicated are two representative plasticity points of Barcelona harbour silts.

Additional necessary information is to know the cyclic stress intensity leading to liquefaction. Different approaches may be found to estimate this stress level. In general, all of them try to estimate the stress ratio (τ/σ'_v: shear stress/vertical effective stress) inducing liquefaction.

Liquefaction is understood as a substantial reduction in undrained strength induced by the accumulation of positive pore water pressures during repeated undrained loading. At the limit, strength reduces to nil values but the back analysis of real cases indicates that some residual shear strength is generally available (Olson and Stark, 2002).

Figure 5.11 shows data originated in earthquake-induced liquefaction cases. It provides the critical stress ratio able to induce liquefaction as a function of the corrected SPT value. It corresponds to an earthquake magnitude of 7.5 and it refers to a reference confining stress of 0.65 atm (65 kPa). In an earthquake of magnitude 7.5, a few strong cycles are applied (15 – 20). This is significantly less than the number of waves hitting the breakwater at maximum storm intensity (H_s = 4 m). But despite the differences between earthquakes and wave loading on vertical caissons, it provides a useful reference value for τ/σ'_v: it may vary between 0.05 and 0.1 if fines content FC < 5% for the range of SPT values (4 – 14) given in Figure 5.7. If FC increases (FC > 35%), the stress ratio increases to 0.1 – 0.2. Cyclic shear tests were performed (reported below) to determine more precisely the stress ratio for liquefaction.

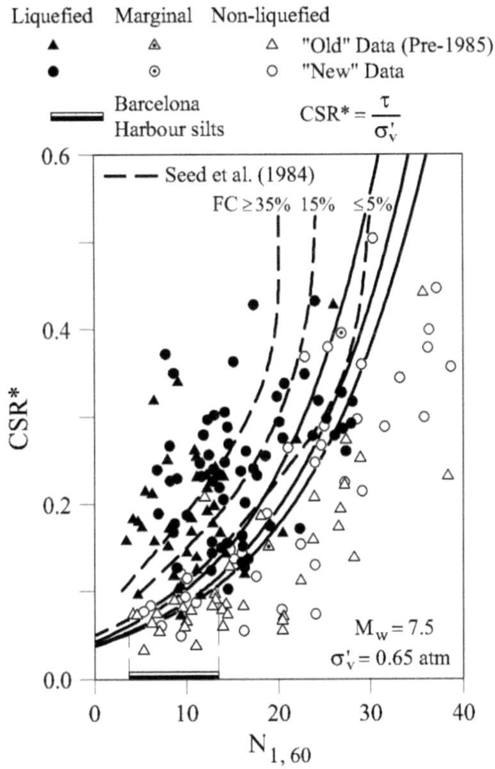

Figure 5.11 Critical stress ratio for liquefaction in terms of SPT values and fine's content (Seed *et al.*, 2003. With permission from ASCE). Also indicated is the range of N_{SPT} values recorded in the soil investigation.

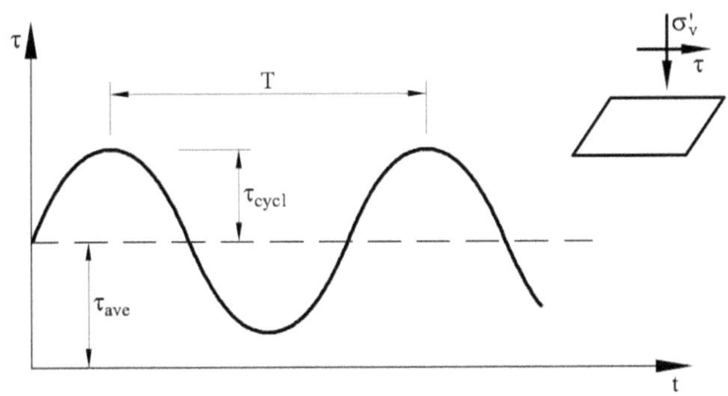

Figure 5.12 Definition of variables in cyclic shear loading.

A related experimental work was performed on undisturbed silt specimens, from the same geological formation, recovered in other Barcelona Harbour emplacements. Undrained simple shear cyclic tests were performed. The cyclic shear excitation is defined in Figure 5.12.

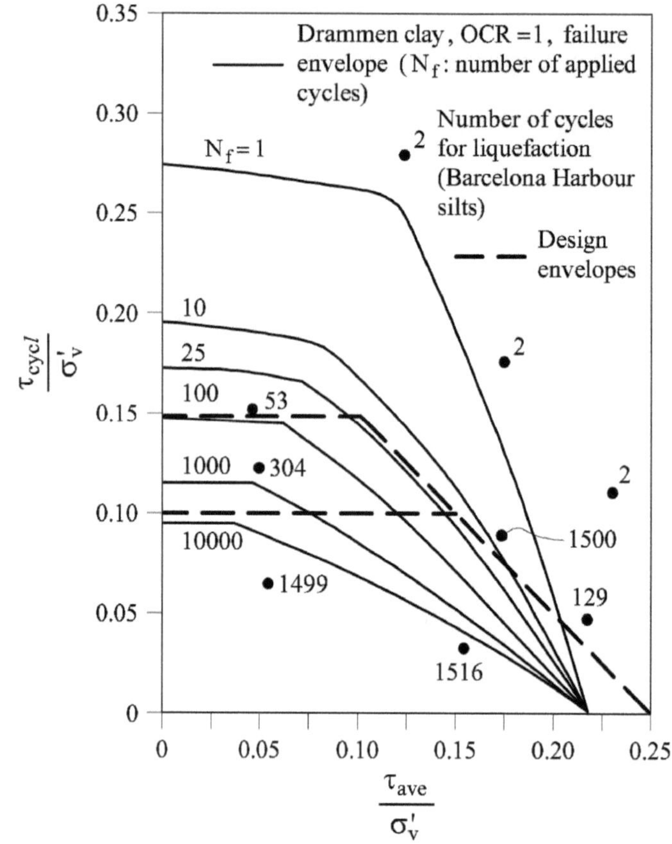

Figure 5.13 Results of cyclic undrained simple shear tests. Also indicated in the figure is data from Drammen clay (NGI 2002).

The periodic shear stress signal is described by an average value, τ_{ave}, a purely cyclic component, τ_{cycl} and a time period *T*. The results of tests performed are represented in a two-dimensional plot relating τ_{cycl}/σ'_v and τ_{ave}/σ'_v (Fig. 5.13). Each of the points in this plot indicates a combination of the pair (τ_{cycl}/σ'_v, τ_{ave}/σ'_v) which leads to failure of the specimen. The number associated with each point is the number of cycles applied. Increasing τ_{cycl}/σ'_v and/or τ_{ave}/σ'_v leads to a progressively smaller number of applied cycles necessary to induce failure. Failure is a consequence of the accumulation of pore pressures which result in increasing shear deformations because of the reduction in normal effective stress acting on the shearing plane. In practice, failure was accepted when the shear strain reached 10%. Also indicated in the plot are the results for Norwegian Drammen clay for normally consolidated conditions.

The information given in the plot may be used to isolate a safe region for a given number of stress cycles applied. Safe regions are limited by the line

$$\frac{\tau_{\text{cycl}}}{\sigma'_v} + \frac{\tau_{\text{ave}}}{\sigma'_v} = \frac{c_u}{\sigma'_v} = a, \tag{5.8}$$

a being defined in Equation (5.1). Equation (5.8) tells that any combination of average and cyclic stress ratios leading to the static strength ratio will lead to failure. When the number of cycles increases, the safe region reduces in size because the cyclic stress ratio decreases. Two safe regions are shown in the plot in Figure 5.13 for a low and a high number of cycles (approximately 40 and 5,000), represented by the upper and lower dashed lines, respectively. In the first case, the limiting cyclic stress ratio is 0.15, provided the average stress ratio does not exceed 0.1. In the second case the limiting cyclic stress ratio is 0.1 for average stress ratios not exceeding 0.15. Beyond this average stress ratio, the cyclic component should be reduced. This plot will be used to estimate liquefaction conditions under wave action in the manner indicated below.

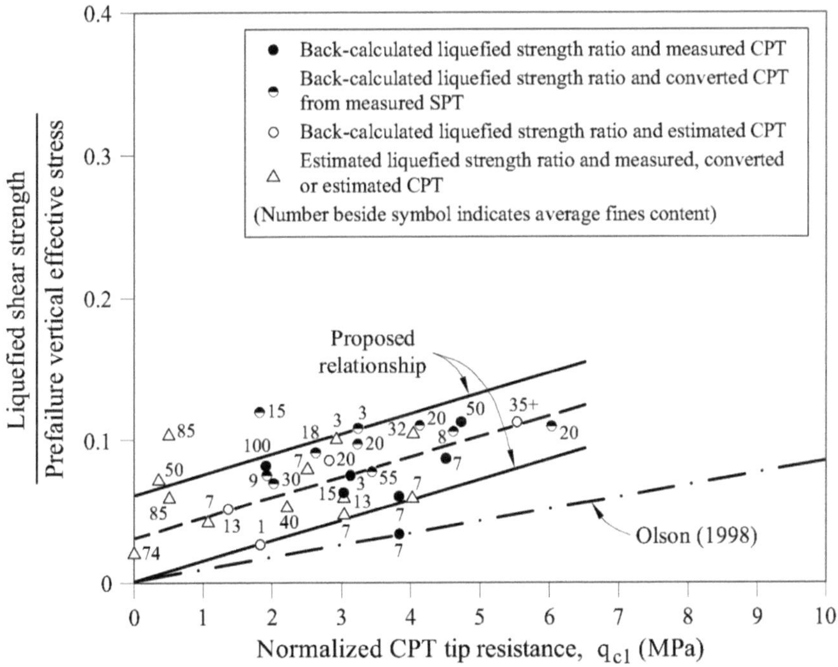

Figure 5.14 Liquefied strength ratio from liquefaction flow failure case histories (Olson and Stark, 2002 © 2008 NRC Canada. Reproduced with permission).

A final point concerns the static strength after liquefaction. In order to answer the question on the strength remaining after liquefaction, static undrained strength tests could be performed after cyclic loading leading to liquefaction. But it is also possible to back analyze some failures involving the flow of soil after cyclic loading. This information is necessary to analyze stability conditions once liquefaction is triggered. Figure 5.14 shows the correlation provided by Olson and Stark (2002) between the liquefied strength ratio and the normalized CPT point resistance. The q_c values reported above for the Barcelona Harbour silty soil, below the caisson foundation, are low (just a fraction of 1 MPa). Figure 5.14 shows that a low value of post-liquefaction strength ratio, ranging between 0 and 0.06, may be operative in this case.

5.4 Settlement Records and Their Interpretation

New caissons, built after the failure, were monitored and settlement records for an extended period of time were obtained (Fig. 5.15). They can be interpreted to derive average values of the foundation soil coefficient consolidation and stiffness. Note that the soil stiffness was already determined in oedometer tests performed on samples. However, the integrated field value provided by caisson settlement records is more reliable.

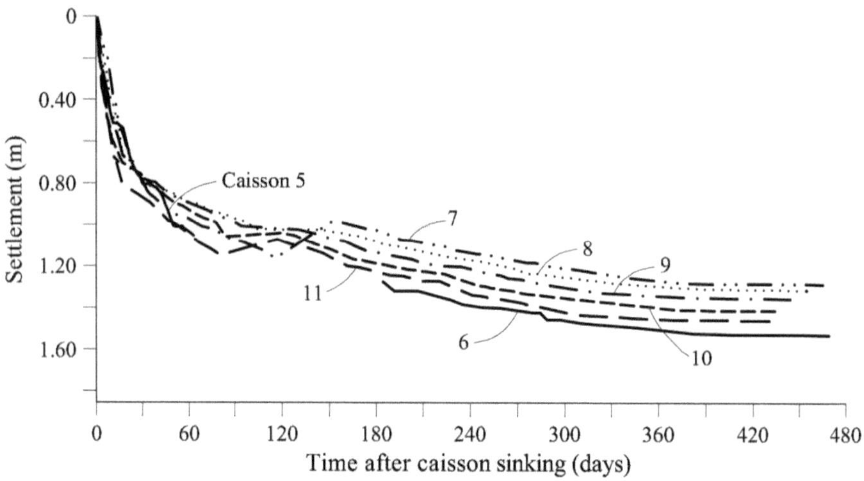

Figure 5.15 Settlement records for Caissons 5, 6, 7, 8, 9, 10 and 11 built after the failure of the first four caissons. Settlement records are plotted with a common time origin.

All the settlement records of caissons located in the vicinity of the failed ones were similar. They could be used to derive a field relationship between degree of consolidation, U, and time. U was calculated, for each time, as the ratio between the current settlement and the maximum value, at long term (around 600 days), which is easily identified in the settlement records. The relationship between U and time is plotted in Figure 5.16 in natural, log scale and square root of time. Settlements are linearly related to the logarithm of time with a good approximation. The root of time plot is non-linear and this is an indication of the progressive reduction in time of the coefficient of consolidation, as shown below.

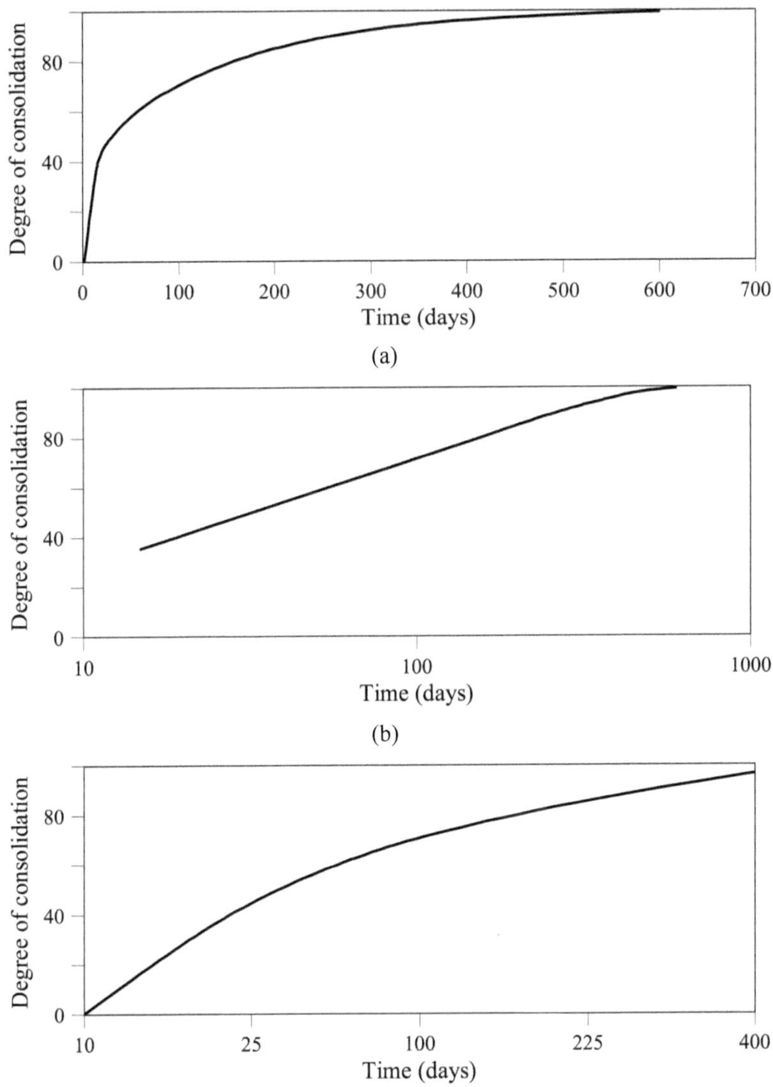

(c)

Figure 5.16 Average degree of consolidation from caisson settlement records: (a) natural time scale; (b) logarithmic time scale; (c) square root of time scale.

Davis and Poulos (1972) published the solution for the consolidation of a strip loading, which is useful to interpret the settlements of a caisson. They made assumptions equivalent to the classical one-dimensional Terzaghi consolidation equation (constant coefficient of compressibility and permeability, soil deforms only in vertical direction). The solution for a permeable top and base is reproduced in Figure 5.17. For a given time, the degree of consolidation increases the more "three-dimensional" is the dissipation effect, i.e. for increasing values of the ratio: thickness of consolidation layer, *h*, over half width of the strip loading, *b*.

In our case, $h/b \approx 2$ and Figure 5.16 indicates that the solution is very close to the one-dimensional Terzaghi solution. The one-dimensional solution is almost exactly reproduced by the closed form equation

$$U = \sqrt{\frac{4T}{\pi}} \tag{5.9}$$

for $U < 0.526$ (see also Chapter 2). Since $T = c_v t / H^2$, Equation (5.9) may be used to find values of c_v (H is the half thickness of the consolidation layer: H = 10.5 m). For every pair of (U, t) values, an estimation of c_v is found. Table 5.1 provides c_v values for the first three months of the consolidation process.

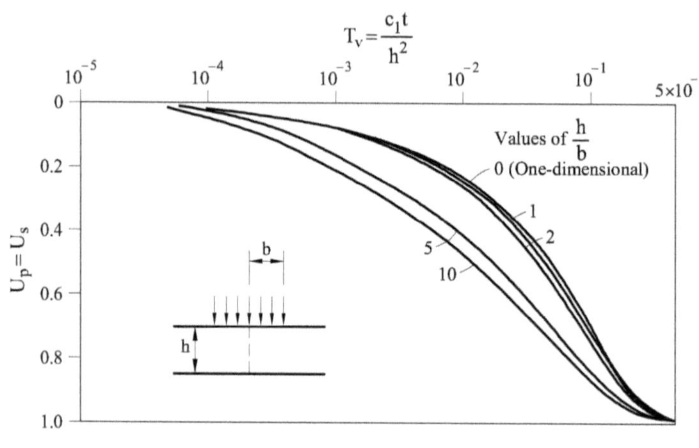

Figure 5.17 Degree of consolidation vs time factor. Strip footing, permeable top, permeable base (after Davis and Poulos, 1972; the original notation has been maintained).

Table 5.1 Coefficient of consolidation from time-settlement records.

t (days)	15	30	45	60	90
c_v (m²/day)	0.748	0.702	0.60	0.53	0.46
c_v (cm²/s)	0.086	0.081	0.07	0.061	0.053

As expected, c_v decreases with time probably because of the reduction in permeability as the soil void ratio decreases. In order to estimate the soil permeability, it is necessary to know the soil confined (elastic) stiffness, E_m, since

$$c_v = \frac{kE_m}{\gamma_w}. \tag{5.10}$$

E_m is expressed in terms of the compressibility index C_c as

$$E_m = \frac{(1+e_0)\sigma'_v}{0.434C_c}. \tag{5.11}$$

Below the caisson base, at a depth equal to caisson half-width, the vertical stress in the soil is estimated to be $(22+10\times8)$ = 300 kPa. Therefore $E_m \approx$ 5,250 kPa. Then, Equation (5.10) provides a soil permeability of $k = 1.5\times10^{-8}$ m/s for c_v = 0.7 m^2/day. For the first stages of consolidation, which are the relevant ones in our case, the foundation soil reacts with a c_v value in the vicinity of 0.75 m^2/day.

The result, even if it is only approximate, indicates that the foundation soil is rather impervious. It will react in an undrained manner when subjected to relatively rapid loading (wave action or caisson sinking) and failure will be also undrained. The relevant strength property will be the undrained strength. Undrained failure will be also the critical one because these soft soils generate positive pore water pressures when sheared, which implies lower strengths, if compared with the drained case (see Fig. 5.18). Given an initial stress state I on the K_0 line, the undrained path (U) will lead to the shear strength c_u. By contrast, a direct application of Coulomb's law implies path D and a higher, unrealistic, and unsafe shear strength τ_f. See Chapter 6 for a more detailed discussion on this issue.

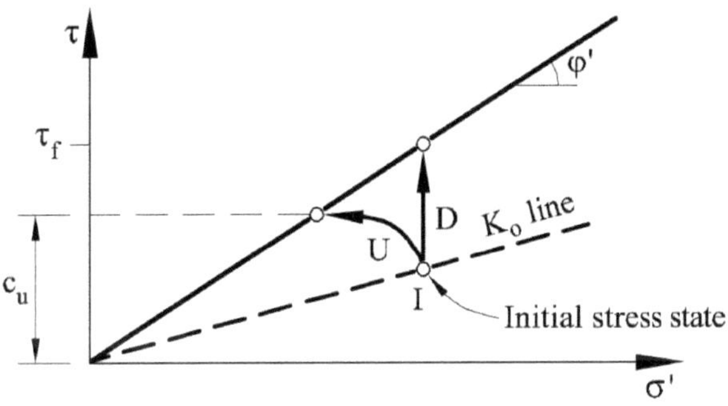

Figure 5.18. Undrained and drained triaxial stress paths.

5.5 Safety During Caisson Sinking

5.5.1 Caisson weight

Once dredging ended in October 2000, the coarse granular base was backfilled into the opened trench on the sea bottom. The berm was levelled and prepared for caisson sinking in May 2001. Actual sinking took place in the middle of October 2001. The storm and the caisson failure arrived 20 days later. (The caisson construction history is schematically indicated in Fig. 5.19.)

Unloading due to dredging and the subsequent granular filling did not restore exactly the initial effective stress in the clayey silt but it was very close. Changes in water content were minor and therefore the natural soil essentially maintained the original undrained strength profile shown in Figure 5.6.

Caissons cell volume amounts to 55% of total volume. When cells are filled with water, the caisson sinks. The granular base was levelled at elevation -17.50 m and, therefore, after sinking, the upper 2 m of caissons remained above the sea level. The effective weight of the caisson per unit length (1 m) in the longitudinal direction is:

$$W'_{\text{water}} = 19.6 \times 19.5 \times (0.55\gamma_w + 0.45\gamma_c) - 19.6 \times 17.5\gamma_w \tag{5.12}$$

where γ_c, γ_w are the unit weights of concrete and sea water. For $\gamma_c = 23\ \text{kN/m}^3$ and $\gamma_w = 10\ \text{kN/m}^3$, $W'_{\text{water}} = 2{,}628\ \text{kN/m}$. The vertical net stress against the foundation is $\sigma_v^{\text{water}} = 2{,}628\ \text{kN/m}/19.6\ \text{m} = 134\ \text{kPa}$.

When filled with submerged sand the caisson effective weight against its foundation, per unit longitudinal length of caisson, is

$$W'_{\text{sand}} = 19.6 \times 19.5 \times (0.55\gamma_{\text{sand}} + 0.45\gamma_c) - 19.6 \times 17.5\gamma_w \tag{5.13}$$

It was estimated that $\gamma_{\text{sand}} = 18\ \text{kN/m}^3$. Also, $\gamma_{\text{concrete}} = 23\ \text{kN/m}^3$ and therefore $W'_{\text{sand}} = Q = 4{,}310$ kN/m. The applied effective vertical stress on the caisson base is $q = Q/19.6$ m = 220 kPa.

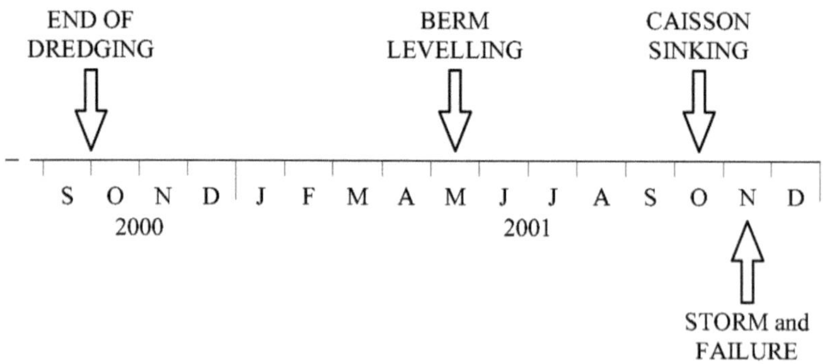

Figure 5.19 History of caisson construction and failure.

5.5.2 Bearing capacity

Davis and Booker (1973) found the exact solution for the bearing capacity of a strip footing when the undrained strength increases linearly with depth according to the relationship

$$c_u = c_{u0} + \rho z, \tag{5.14}$$

where ρ is a constant. The upper granular layer (Fig. 5.6) introduces some additional bearing capacity, but it is probably very small. In fact, the lateral extent of the berm is small. If a simple failure mechanism is considered, the sole effect of the granular layer is to provide a frictional resistance T (Fig. 5.20). Being at the surface and having only a thickness of around 2 m, the confining stress in the granular berm is very small compared with the contribution of the natural soil. It will be accepted, for simplicity, that only the natural soil contributed to the bearing capacity.

The theoretical bearing capacity was written, by Davis and Booker (1973),

$$Q/B = F\left[(\pi+2)c_{u0} + \rho b/4\right], \tag{5.15}$$

where F is a correction factor that depends on the ratio $\rho b/c_{u0}$ and may be found in Figure 5.21.

Equation (5.15), for c_{u0} = 20.25 kPa; $\rho = (67.5-20.25)\,\text{kPa}/21\,\text{m} = 2.25\,\text{kN/m}^3$ and F = 1.35 (for $\rho b/c_{u0}$ = $2.25\,\text{kN/m}^3 \times 19.6\,\text{m}/20.25\,\text{kN/m}^2 = 2.2$; see Figure 5.20 for rough footing), provides $Q/b = 155$ kPa, which is lower than the applied caisson net stress if filled with sand (220 kPa).

Caissons, however, were initially filled with water before replacing it with sand. The net stress on the foundation of a water filled caisson is 134 kPa. Therefore, the safety factor when the caissons were sunk with water ballast can be calculated as $SF = 155/134 = 1.16$.

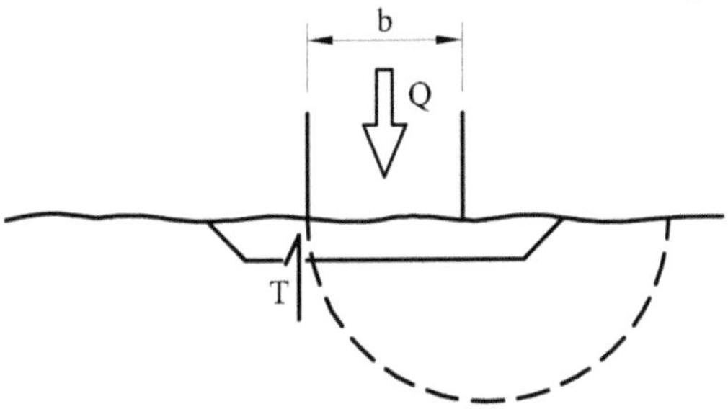

Figure 5.20 Sketch to illustrate the frictional resistance to failure offered by the granular berm.

The theoretical expression (5.15) predicts conditions very close to failure at the time of caisson sinking. Three-dimensional effects, due to the finite rectangular shape of the caissons base, leads to an increase of the bearing capacity and this effect may explain that the caissons remained stable. Also, the limited thickness of the upper granular term provided some additional bearing capacity. On the other hand, if the undrained strength was actually somewhat higher (Eq. (5.1) with a = 0.30, for instance), the safety factor against failure would also increase. Safety factor increases linearly with c_{u0} and a value of a = 0.30 would result in $SF = 186/134 = 1.4$.

It is difficult to be more precise, but the fact is that the caissons did not fail during sinking. However, the estimation made points towards a small safety factor above one. Beyond this moment, the consolidation of the soil under the caisson's weight will increase the available shear strength in the natural soil. The caisson was then loaded with its definite weight (when filled with saturated sand), it consolidated during a few weeks, and eventually it received the storm-wave loading.

Before analyzing these processes, it is worth investigating if the theoretical undrained bearing capacity obtained by Davis and Booker for a strip footing on a clay soil whose strength increases with depth, may be approximated by a simpler kinematically admissible mechanism, using the upper bound theorem of plasticity. The reason behind this approach is to prepare the ground for subsequent calculations involving the wave action and a more complex distribution of undrained strength with depth. In fact, the consolidation process will lead to a "map" of c_u values which will match the "map" of mean effective stresses (Eq. (5.2)). This distribution, changing with time, will be substantially different from a linear variation of c_u and it will not be amenable to theoretical solutions. However, the plasticity theorems still provide an approximation. It then seems wise to try to establish some confidence on the assumed failure mechanism (based on the upper bound theorem). The way to do it is by comparing the exact solution provided by Davis and Booker (1973) with the upper bound approximation.

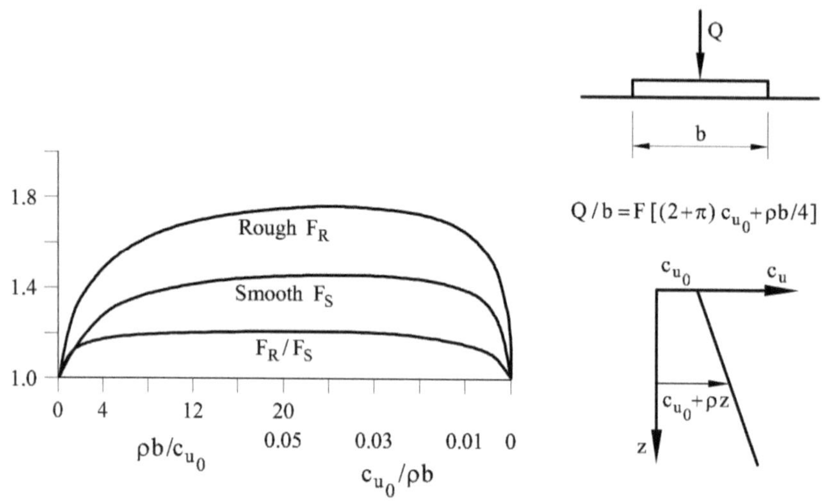

Figure 5.21 Correction factors for rough and smooth footings (after Davis and Booker, 1973).

5.5.3 An upper bound solution for a rough strip footing founded on clay with a linearly increasing strength with depth

Figure 5.22 shows the critical velocity field found by Davis and Booker (1973) in their exact solution to the bearing capacity problem (infinite, rigid smooth footing). The mechanism is symmetric with respect to the axis of the footing.

Even if the rough footing is a more realistic case, this mechanism suggests the simplified symmetric mechanism, based on rigid triangular wedges, indicated in Figure 5.23.

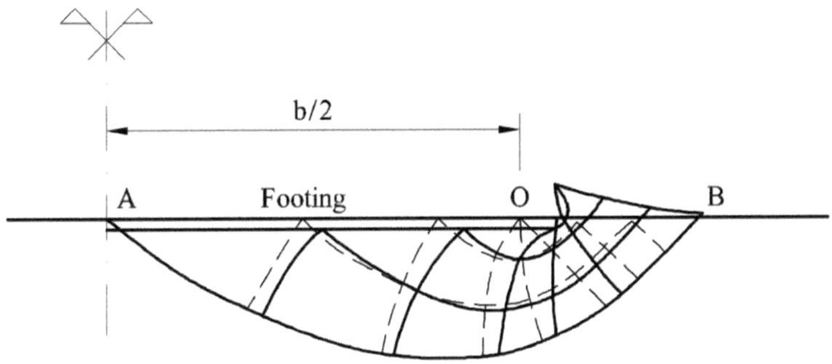

Figure 5.22 Velocity field for smooth footing (after Davis and Booker, 1973).

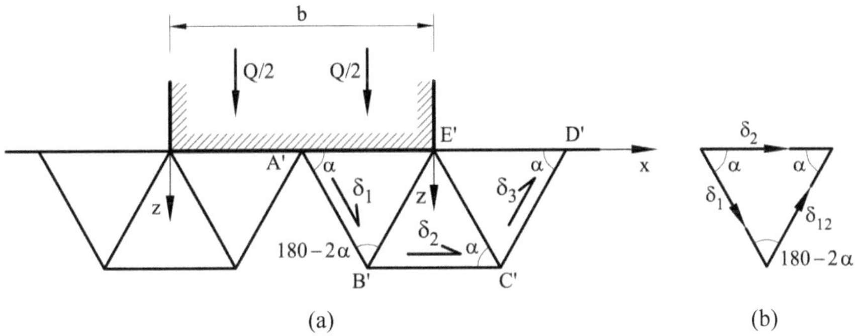

Figure 5.23 Symmetric failure mechanism for upper bound analysis of strip footing under vertical load.

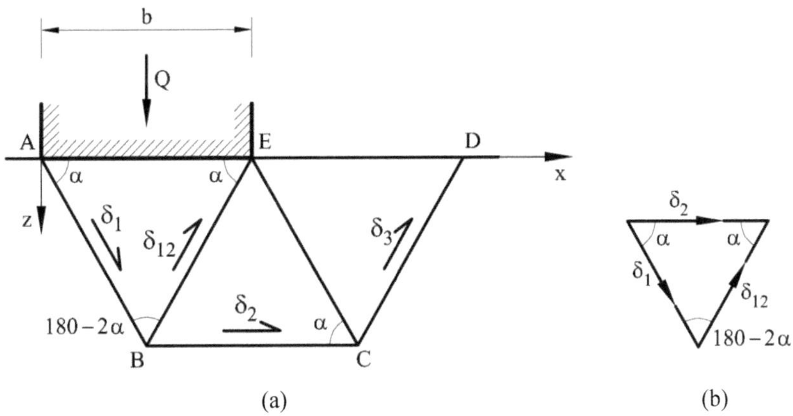

Figure 5.24 Nonsymmetric failure mechanism for upper bound analysis of strip footing under vertical load.

Consider one of the two symmetric mechanisms that receive half of the external load ($Q/2$). This mechanism will be optimized with respect to the angle α

shown in Figure 5.23. The motion of this mechanism is characterized by the virtual displacement rate vector δ_1, which defines the motion of wedge $\mathrm{A'B'E'}$ sliding on the straight segment $\mathrm{A'B'}$. The length of segment $\mathrm{A'B'}$ in terms of caisson width, *b*, is $L = b/(4\cos\alpha)$. The undrained strength will be defined by the linear function $c_u(z) = c_{u0} + \rho z$.

The dissipation along $\mathrm{A'B'}$ is calculated as follows

$$
\begin{aligned}
W_{\mathrm{A'B'}} &= \int_{\mathrm{A'}}^{\mathrm{B'}} c_u(z)\delta_1 \mathrm{d}l = \int_{\mathrm{A'}}^{\mathrm{B'}} (c_{u0} + \rho z)\delta_1 \mathrm{d}l = \int_{\mathrm{A'}}^{\mathrm{B'}} (c_{u0} + \rho l \sin\alpha)\delta_1 \mathrm{d}l = \\
&= \delta_1 \left[c_{u0} l + \frac{\rho}{2} l^2 \sin\alpha \right]_0^L = \delta_1 \frac{b}{4\cos\alpha}\left(c_{u0} + \frac{\rho b}{8}\tan\alpha \right).
\end{aligned}
\tag{5.16}
$$

Consider now the dissipation along $\mathrm{B'C'}$. The relative motion δ_2 between the moving wedges $\mathrm{B'C'E'}$ and the rigid soil along $\mathrm{B'C'}$ is given by $\delta_2 = 2\delta_1\cos\alpha$ in view of the motion compatibility condition expressed in Figure 5.23b.

The *z* coordinate of line $\mathrm{B'C'}$ is $z = b\tan\alpha/4$; then

$$
W_{\mathrm{B'C'}} = c_u\left(z = \frac{b\tan\alpha}{4} \right)\delta_2 \frac{b}{2} = \left(c_{u0} b\cos\alpha + \rho\frac{b^2}{4}\sin\alpha \right)\delta_1 . \tag{5.17}
$$

Considering the hypothesis of rough footing, dissipation along the line $\mathrm{A'E'}$ is

$$
W_{\mathrm{A'E'}} = c_{u0}\frac{b}{2}\delta_1 \cos\alpha . \tag{5.18}
$$

Dissipation along lines $\mathrm{A'B'}$, $\mathrm{B'E'}$, $\mathrm{E'C'}$ and $\mathrm{C'D'}$ is equal and therefore the total internal dissipation work on the mechanism will be

$$
W_{\mathrm{int}} = 4W_{\mathrm{A'B'}} + W_{\mathrm{B'C'}} + W_{\mathrm{A'E'}} . \tag{5.19}
$$

The external work performed by $Q/2$ is calculated as

$$
W_{\mathrm{ext}} = \frac{Q}{2}\delta_1 \sin\alpha . \tag{5.20}
$$

Making W_{int}=W_{ext}, Q is isolated in terms of α, c_{u0}, ρ, b:

$$
Q = \frac{b}{4}\frac{\left(8c_{u0} + \rho b\tan\alpha + 12 c_{u0}\cos^2\alpha + 2\rho b\sin\alpha\cos\alpha \right)}{\sin\alpha\cos\alpha} . \tag{5.21}
$$

The best upper bound solution for Q is its minimum value with respect to α. This minimization calculation was performed with the help of the built in "solve" function including in Excel. For b = 19.6 m, c_{u0} = 20.25 kPa, and ρ = 2.25 kN/m^3 a minimum value of Q/b = 182 kPa is obtained for a critical angle α equal to 50.8°.

The theoretical value (Davis and Booker, 1973) for rough footing provides a value Q/b = 155 kPa. The error of the simple upper bound mechanism is 17%, a reasonable value in practical terms. This result indicates that the triangular wedge mechanism in Figure 5.22 is an acceptable approximation to calculate the bearing capacity factor for strip footings resting on a clay soil with a linearly increasing strength with depth. It should be stressed that the main purpose of the analysis developed in the remaining part of the chapter is to examine the variation of safety factor during the consolidation after caisson sinking, subsequent filling with sand, and storm action. The upper bound calculations provide a simple and practical tool to evaluate the sequence of events leading to caisson failure.

However, if a horizontal load is also acting on the caisson, because of wave action, a symmetric mechanism cannot possibly occur and a more likely mechanism is indicated in Figure 5.24. We call it a nonsymmetric failure mechanism. Let us consider first this mechanism under a vertical load. The calculation is now almost identical to the previous one. Note that in this case the caisson will not displace with respect to the wedge AEB along the line AE. Therefore, no dissipation will be calculated on segment AE. Repeating previous steps, the dissipation on segments AE and BC will be

$$W_{\text{AB}} = \frac{\delta_1 b}{2\cos\alpha}\left(c_{u0} + \frac{1}{4}\rho b\tan\alpha\right), \tag{5.22}$$

$$W_{\text{BC}} = \left(2bc_{u0}\cos\alpha + \rho b^2\sin\alpha\right)\delta_1. \tag{5.23}$$

The internal dissipation work will be $W_{\text{int}} = 4W_{AB} + W_{BC}$. The external work is now given by:

$$W_{\text{ext}} = Q\delta_1\sin\alpha. \tag{5.24}$$

Making the two works equal and isolating Q,

$$Q = \frac{b}{2}\frac{\left(4c_{u0} + \rho b\tan\alpha + 4c_{u0}\cos^2\alpha + 2\rho b\sin\alpha\cos\alpha\right)}{\sin\alpha\cos\alpha}, \tag{5.25}$$

which is different from the Q value calculated for the symmetric mechanism.

The minimization of Q with respect to α was also performed on an Excel sheet. For the same parameters previously considered (b = 19.6 m; c_{u0} = 20.25 kPa; ρ = 2.25 kN/m^3), a failure unit load Q/b = 209 kPa was calculated (for a critical angle α equal to 44.5°) which is 15% higher than the load calculated for the symmetric mechanism. This nonsymmetric mechanism will be used when wave action is considered, as mentioned above.

Caisson consolidation resulted in increasing c_u values and in increasing safety factor against bearing capacity failure.

Two weeks after the caisson first sinking, caisson cells were filled with sand. This increase in net weight will be also analyzed below in order to determine the associated safety factor. In the days that followed the sand filling, caisson consolidation continued and the soil undrained strength had to be estimated before analyzing the storm effect on caissons.

Therefore, the next step in the analysis was to investigate the caisson consolidation and the increase in soil strength.

5.6 Caisson Consolidation. Increase in Soil Strength

Actual settlement records of caissons built after the failure, shown in Figure 5.15, indicate that a significant consolidation may be achieved in a few weeks. Points of the soft foundation soil close to the upper pervious granular layer would experience a rapid consolidation under the full caisson loading. Clay levels located close to the lower pervious sandy boundary will also consolidate fast but the stress increments reaching the lower sand levels will be significantly lower. Caisson consolidation leads to a progressive increase in effective stress and therefore to an increase in undrained strength. For the reasons mentioned, however, the new distribution of undrained strength values will be non-homogeneous and also far from the initial linear distribution with depth. This will be especially the case of the foundation soil directly under the caissons.

The increments of undrained strength will be simply calculated as a fraction of the increment in effective mean stress through Equation (5.2). Therefore, the objective now is to calculate the distribution of effective mean stress under the caisson loading, taking into account the consolidation process. The calculation will be split into two parts:

- Stress increments under a strip footing and determination of excess pore pressures.
- Dissipation of the induced excess pore pressures.

5.6.1 Stress increments under a strip footing and determination of excess pore pressures

This analysis will be guided by the subsequent use of the calculated undrained strengths. In fact, the ultimate objective is to determine the failure load and to compare it with the actual caisson loading. Failure conditions will be calculated by means of the upper bound theorem of plasticity, through the mechanisms already examined (Figs. 5.23 and 5.24). Consider in Figure 5.25 the two alternative mechanisms proposed here: a symmetric one (already identified as an adequate solution for vertical loading only) and the non-symmetric, one which will be employed when including wave action.

Upper bound calculation will require the determination of plastic work dissipation on segments AB, A′B′, etc. Since non-linear strength variations will be the rule, a minimum of three control points are proposed to estimate by a simple numerical integration, the average strength on each of the sliding surfaces of the two mechanisms shown in the figure. Those points (marked as open circles) define a number of vertical profiles characterized by the horizontal coordinate distances to the left caisson foundation corner indicated in Figure 5.25.

Pore pressure dissipation will be dominated by the vertical flow towards the upper and lower drained boundaries. A hypothesis of vertical consolidation, which is close to real conditions, as justified in Equation (5.4), helps to perform the consolidation analysis. The following sequence of steps will be considered (the starting point – time = 0 – will be the initial sinking of caissons):

a) Stress increments in the foundations are determined. Elastic solutions for strip loading are used. Stress calculations are performed in vertical profiles located on the horizontal coordinates shown in Figure 5.25 (x = 0, 4.9, 9.8 m… etc.). The reason for this choice has already been given.
b) Mean total stress will be computed in points located on the vertical profiles. Excess pore pressures (over hydrostatic values) will be made equal to the increments of mean stress. This is a reasonable and sufficiently accurate assumption.
c) Excess pore pressures will be dissipated vertically towards the upper and lower drainage boundaries in a one-dimensional process. A time period of 14 days will provide the state of the foundation before sand filling the caissons.
d) Mean effective stress will be calculated as a difference between calculated total stresses and pore water pressures. An updated distribution of c_u values will be calculated through Equation (5.2).
e) Caisson failure loads will be determined through the upper bound theorem. The safety factor will be determined.

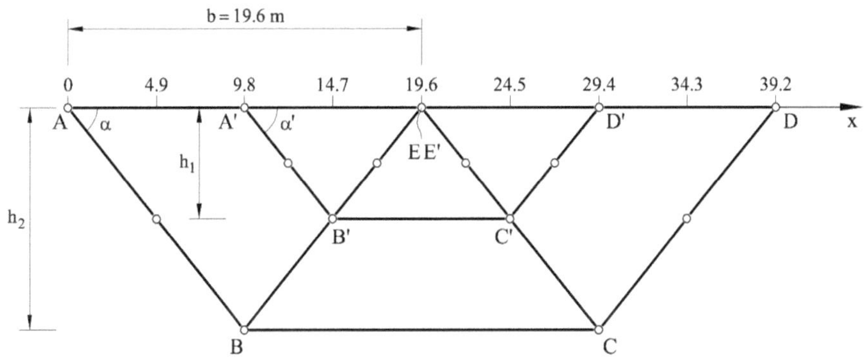

Figure 5.25 Geometry of the symmetric (A′B′C′D′E′) and nonsymmetric (ABCDE) failure mechanisms. Position of vertical profiles for the calculation of undrained soil strength.

5.6.2 Stress increments

a) Calculation of stress increments in the foundation

Poulos and Davis (1973) published the stress distribution beneath a strip loading uniformly loaded. With reference to Figure 5.26, the stresses on a point in an elastic half space, defined by their coordinates (x,z) or, alternatively, by angles α and δ are given by

$$\sigma_z = \frac{q}{\pi}\left[\alpha + \sin\alpha\cos\left(\alpha + 2\delta\right)\right], \tag{5.26a}$$

$$\sigma_x = \frac{q}{\pi}\left[\alpha - \sin\alpha\cos\left(\alpha + 2\delta\right)\right], \tag{5.26b}$$

$$\sigma_y = \nu\left(\sigma_x + \sigma_z\right), \tag{5.26c}$$

$$\tau_{xz} = \frac{q}{\pi}\sin\alpha\cos\left(\alpha + 2\delta\right). \tag{5.26d}$$

b) Mean stress and initial excess pore pressures

The mean stress is calculated as

$$\sigma_m = p = \frac{\sigma_x + \sigma_y + \sigma_z}{3} = \frac{2(1+\nu)}{3}\frac{q}{\pi}\alpha, \tag{5.27}$$

where

$$\alpha = \arctan\left(\frac{x}{z}\right) + \arctan\left(\frac{b-x}{z}\right) \tag{5.28}$$

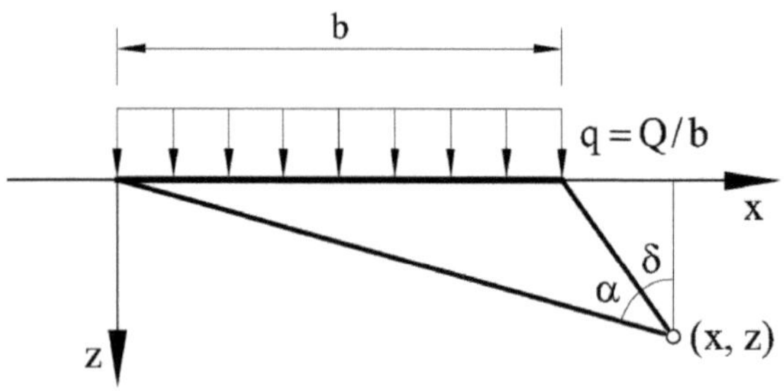

Figure 5.26 Uniform strip loading. Coordinate system (Poulos and Davis, 1973).

5.6.3 Initial excess pore pressures

The calculated profiles of σ_m (which is equal to the initial excess pore pressure) at a few horizontal coordinates indicated in Figure 5.25 (x = 9.8, 12.25, 19.6, 22.05 26.95 and 34.30 m) are given in Figure 5.28 for t = 0. The actual sequence of caisson loading is shown in Figure 5.27. After sinking, caissons remained full of water during 14 days. The external load q in this period was the net stress applied by the caisson at the time of sinking (caissons filled with water: $q = Q/b$ = 134 kPa). The time of sinking is t = 0 for the remaining of the analysis. At t = 14 days caisson cells were filled with sand and the net stress on foundation increased to 220 kPa. The storm arrived at t = 21 days.

Figure 5.28 shows two kinds of shapes for the distribution of the initial excess pore pressure due to the caisson loading (solid line). Under the caisson ($x = 0^+$ to

$x = 19.6$ m), the mean stress reaches a maximum at the caisson-soil contact and a minimum at the bottom of the clay stratum (at a depth of 21 m under the caisson's base, where the pervious sand layer is encountered). On both sides of the caisson ($x > 19.6$ m) the stress increment at the surface is zero. It increases to reach a maximum at some intermediate depth and decreases again. The intensity of the mean stress increments decreases as the distance to the caisson base increases.

A dissipation process of excess pore pressures will immediately start towards the upper and lower drainage boundaries and, in parallel, mean effective stresses will increase. These are steps c) and d) of the description of the process leading to an increase in soil strength.

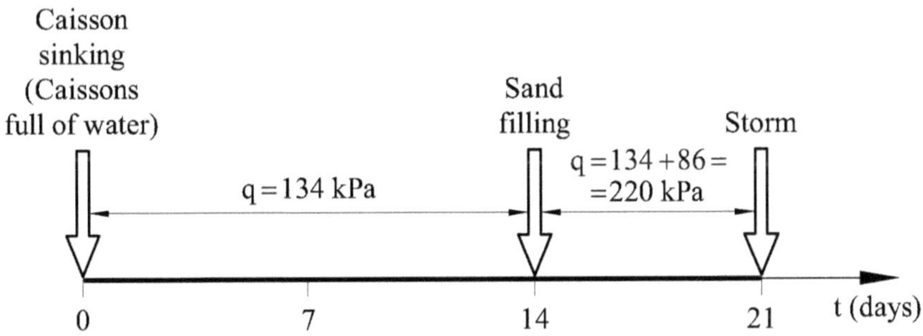

Figure 5.27 Caisson loading sequence.

5.6.4 Excess pore pressure dissipation

The initial excess pore pressures plotted in Figure 5.28 do not follow simple linear laws. This prevents the direct use of consolidation solutions given in most soil mechanics textbooks. Probably the simplest curve fitting all the profiles shown in the figure to a reasonable approximation is a parabola:

$$f(z) = A + Bz + Cz^2, \tag{5.29}$$

where $f(z)$ is the initial excess pore pressure, which is a function of the vertical coordinate. *A, B, C* can be determined by curve fitting procedures or simply by selecting three points on the curve at different *z* to be fitted. Then, the coefficients *A, B, C* are derived from the following system of algebraic equations:

$$f(z_1) = A + Bz_1 + Cz_1^2, \tag{5.30a}$$

$$f(z_2) = A + Bz_2 + Cz_2^2, \tag{5.30b}$$

$$f(z_3) = A + Bz_3 + Cz_3^2. \tag{5.30c}$$

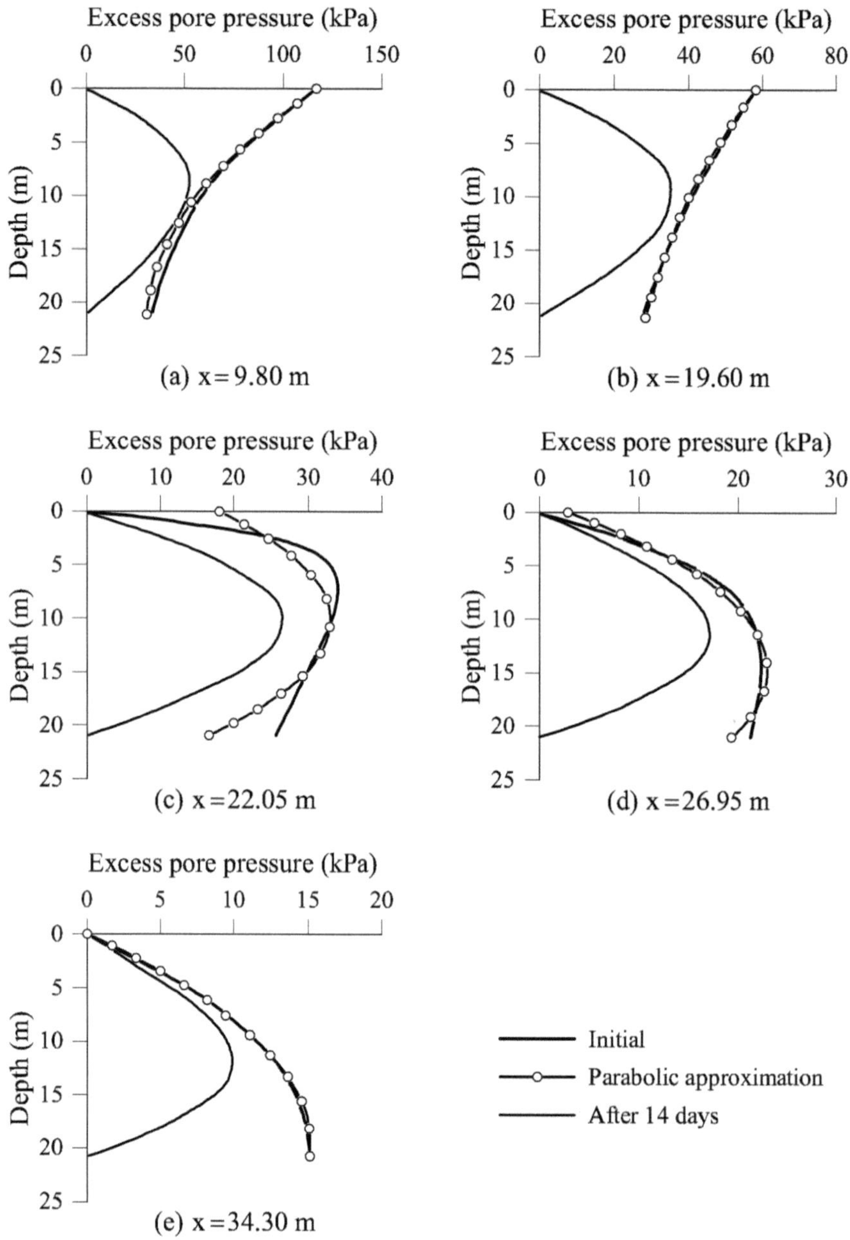

Figure 5.28 Excess pore pressures at the vertical profiles defined by x = 9.8, 19.6, 22.05, 26.95 and 34.30 m in Figure 5.25. Each one of the plots provides the calculated initial excess pore pressure, its parabolic approximation, and the excess pore pressure profile calculated after 14 days of consolidation.

The "solve" function of the Maple program provides the following solution:

$$A = \frac{f(z_1)\left[z_2^2 z_3 - z_2 z_3^2\right] + f(z_2)\left[z_1 z_3^2 - z_1^2 z_3\right] + f(z_3)\left[z_1^2 z_2 - z_1 z_2^2\right]}{z_2 z_1^2 + z_3 z_2^2 + z_1 z_3^2 - z_3 z_1^2 - z_1 z_2^2 - z_2 z_3^2}, \tag{5.31a}$$

$$B = -\frac{z_1^2\left[f(z_3) - f(z_2)\right] + z_2^2\left[f(z_1) - f(z_3)\right] + z_3^2\left[f(z_2) - f(z_1)\right]}{z_2 z_1^2 + z_3 z_2^2 + z_1 z_3^2 - z_3 z_1^2 - z_1 z_2^2 - z_2 z_3^2}, \tag{5.31b}$$

$$C = \frac{z_1\left[f(z_3) - f(z_2)\right] + z_2\left[f(z_1) - f(z_3)\right] + z_3\left[f(z_2) - f(z_1)\right]}{z_2 z_1^2 + z_3 z_2^2 + z_1 z_3^2 - z_3 z_1^2 - z_1 z_2^2 - z_2 z_3^2}. \tag{5.31c}$$

The preceding equations were introduced in an Excel sheet and the fit of the excess pore pressure profiles was obtained by selecting points of the theoretical distribution of Δu (=Δp). The selection of fitting points was directed towards a faithful representation of pore pressures in the central zone of the consolidating layer. The reason is that the excess pore pressures at the upper and lower boundaries will become zero (this is the boundary condition) immediately after the start of the consolidation process. In other words, the precise representation of the extreme values of excess pore pressures is not so relevant. Of course, the total area of the excess pore pressures should be maintained equivalent in the theoretical and the fitted initial pore pressure profile.

The result of this fitting process is also given in Figure 5.28. The parabolic fit is very good under the caisson and at a certain distance from it. The most difficult fit is for vertical profiles outside the caisson but in its immediate vicinity. However, increasing the degree of the polynomial approximation was probably not a reasonable decision in view of the associated complexity and the limited influence of an exact representation of initial excess pore pressures.

The excess pore pressures will be assumed to dissipate vertically, as mentioned before. The problem is schematically shown in Figure 5.29. The excess pore pressure $u(z,t)$ must satisfy Terzaghi's classical equation

$$c_v \frac{\partial^2 u}{\partial z^2} = \frac{\partial u}{\partial t}, \tag{5.32}$$

where c_v is the coefficient of consolidation subjected to the following boundary and initial conditions:

Boundary condition ($\forall t$):

$$u(z = 0, t) = 0, \tag{5.33a}$$

$$u(z = 2H = 21 \text{ m}, t) = 0. \tag{5.33b}$$

Initial condition ($\forall z$):

$$u(z,t=0) = A + Bz + Cz^2, \tag{5.34}$$

where coefficients A, B and C are determined through Equations (5.31a,b,c) for each of the vertical excess pore pressure profiles defined in Figure 5.25.

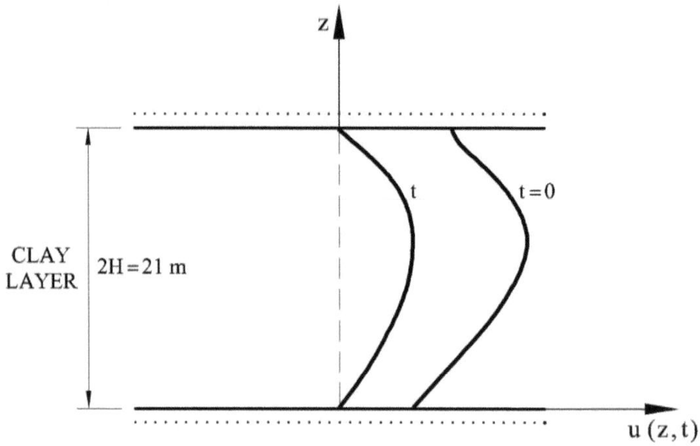

Figure 5.29 Double-drained consolidation induced by caisson sinking at t = 0 and excess pore pressures.

Note that the similarity of excess pore pressures under the caisson will lead to similar values of coefficients A, B, C. Therefore, no significant pore pressure gradients in the horizontal direction will occur under the central part of the caisson foundations. A one-dimensional vertical dissipation is close to real conditions. In the vicinity of the edges (compare pore pressure profiles for x = 19.6 m and x = 22.95 m in Fig. 5.28), horizontal pore pressures gradients are higher, however. This will lead to some horizontal flow components which are not taken into account. At increasing distance from the caisson (pore pressure profiles at x = 26.95, 34.30 m in Fig. 5.28) they again become similar among them and the vertical dissipation dominates the process. Recall also that the solution of the two-dimensional consolidation problem (in terms of the degree of consolidation) was very close to the one-dimensional case for the geometry of the caisson and its foundation (Fig. 5.17).

In terms of the dimensionless variables $Z = z/H$; $T = t/\tau$ and $W = u/u_0$, where τ and u_0 are a reference time $\left(\tau = H^2/c_v\right)$ and a reference pressure, the consolidation equation becomes

$$\frac{\partial^2 W}{\partial Z^2} = \frac{\partial W}{\partial T}. \tag{5.35}$$

The boundary and initial conditions become

Boundary condition ($\forall T$):

$$W\left(Z=0,T\right)=0, \tag{5.36a}$$

$$W\left(Z=2,T\right)=0. \tag{5.36b}$$

Initial condition ($\forall Z$):

$$W\left(Z,T=0\right)=A+BHZ+CH^2Z^2=\bar{A}+\bar{B}Z+\bar{C}Z^2, \tag{5.37}$$

where a new set of constants ($\bar{A}=A$, $\bar{B}=BH$ and $\bar{C}=CH^2$) are defined to describe the initial parabolic excess pore pressure in terms of the dimensionless coordinate Z.

The general solution of Equations (5.35), (5.36) and (5.37) is (Alonso and Krizek, 1975)

$$W\left(Z,T\right)=\int_R g\left(Z,T/Z_0,0\right)f\left(Z_0\right)\mathrm{d}Z_0-\int_0^T \mathrm{d}T_0\int_S \frac{\partial g\left(Z,T/Z_0,T_0\right)}{\partial n_0}W\left(Z_0,T_0\right)\mathrm{d}S_0, \tag{5.38}$$

where R is the domain of integration ($Z=0$ to $Z=2$ in our case); S is the boundary of R ($Z=0;\ Z=2$); n_0 is the normal to the boundary (the Z direction); $W(Z_0,T_0)$ are the boundary conditions (homogeneous in the present case), and g is the Green function associated with the consolidation equation.

Function $g\left(Z,T/Z_0,T_0\right)$ is the solution of the consolidation equation, in the case of homogeneous boundary conditions, when a unit "impulse" increase in pore pressure is introduced in coordinate $Z=Z_0$ at time $T=T_0$. The solution to this problem is

$$g\left(Z,T/Z_0,T_0\right)=\sum_{n=1}^{\infty}\sin\left(\frac{n\pi Z_0}{2}\right)\sin\left(\frac{n\pi Z}{2}\right)\exp\left(\frac{-n^2\pi^2}{4}\left(T-T_0\right)\right). \tag{5.39}$$

In Equation (5.38), $f(Z_0)$ is the initial condition (the parabolic distribution of pore pressures in our case).

Therefore, taking Equations (5.36) to (5.39) into account,

$$\begin{aligned}W\left(Z,T\right)&=\int_0^2\left(\sum_{n=1}^{\infty}\sin\left(\frac{n\pi Z_0}{2}\right)\sin\left(\frac{n\pi Z}{e}\right)\exp\left(-\frac{n^2\pi^2T}{4}\right)\right)\left(\bar{A}+\bar{B}Z_0+\bar{C}Z_0^2\right)\mathrm{d}Z_0=\\&=\sum_{n=1}^{\infty}\sin\left(\frac{n\pi Z}{2}\right)\exp\left(-\frac{n^2\pi^2T}{4}\right)\int_0^2\left[\bar{A}\sin\left(\frac{n\pi Z_0}{2}\right)+\bar{B}Z_0\sin\left(\frac{n\pi Z_0}{2}\right)+\bar{C}Z_0^2\sin\left(\frac{n\pi Z_0}{2}\right)\right]\mathrm{d}Z_0=\\&=\sum_{n=1}^{\infty}\frac{2}{n\pi}\sin\left(\frac{n\pi Z}{e}\right)\exp\left(-\frac{n^2\pi^2T}{4}\right)\left\{\bar{A}\left[(1)^{n+1}+1\right]+2\bar{B}(-1)^{n+1}+4\bar{C}\left[\frac{2}{n^2\pi^2}\left[(-1)^n-1\right]-(-1)^n\right]\right\}\end{aligned} \tag{5.40}$$

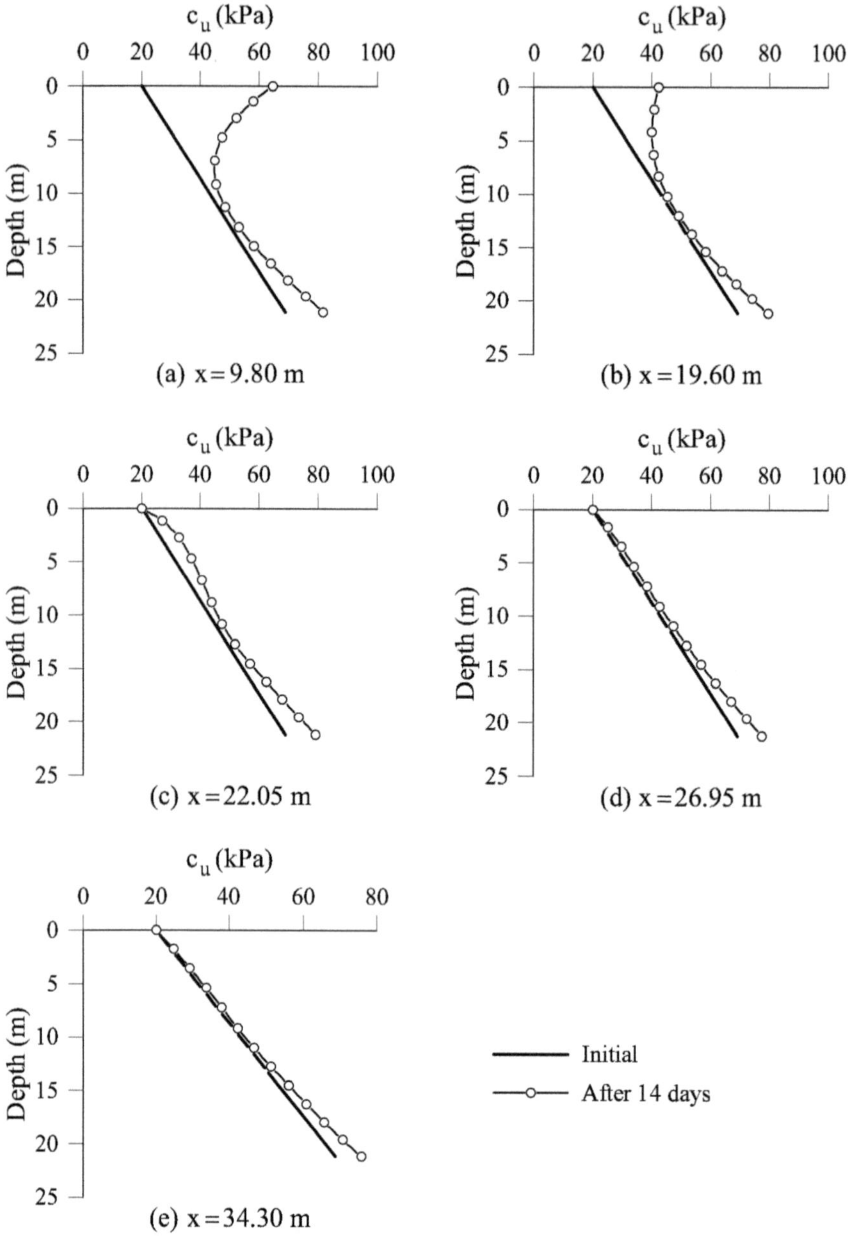

Figure 5.30 Undrained strength at the vertical profiles defined by x = 9.8, 19.6, 22.05, 26.95 and 34.30 m in Figure 5.25. Each one of the plots provides the calculated initial strength and the resulting profile 14 days after the beginning of consolidation, when caissons were still filled with water.

5.6.5 Effective stresses and updated undrained strength

Equation (5.40) was used to calculate the excess pore pressure 14 days after caisson sinking (when full of water). Adding the first five terms in Equation (5.40) provides almost the exact solution. The calculated pore water pressures are also plotted in the vertical profiles given in Figure 5.28. Fourteen days after caisson sinking, the excess pore pressure in all profiles has a similar shape with maximum values at depths ranging between 7 and 13 m. The increase in mean effective stress is calculated as

$$\Delta\sigma'_m = \Delta\sigma_m - \Delta u, \tag{5.41}$$

where $\Delta\sigma_m$ is the increment in total stress (Eq. (5.27) and profiles for $t = 0$ in Fig. 5.28) and Δu is calculated through Equation (5.40). All the necessary information is included in the plots of Figure 5.28.

The updated effective stress at t = 14 days is calculated by adding the initial stress state and the increment provided by Equation (5.41).

The new c_u profiles, calculated through $c_u \approx 0.25\sigma_v$, are compared with the initial values in Figure 5.30. A significant increase in c_u occurs in the first 5 m under the caisson. A smaller increase close to the lower drained boundary has a negligible effect on the increase in caisson failure load. Note also that beyond the loaded area (x > 19.60 m), the increase in undrained strength is very small at any depth.

The new distribution of c_u values under the caisson is now fundamentally different if compared with the initial linear increase with depth. Now c_u reaches a local maximum directly under the caisson and decreases with depth until reaching the initial distribution of undrained strength, which increases linearly with depth. This change will have interesting effects on the critical failure mechanism, as explained below.

The next step in caisson construction was increasing caisson loading to its full design value (sand filling). The safety factor against failure will now be calculated as well as the subsequent consolidation process under the new load. This will lead to an updated distribution of undrained strength, which will be operating at the time of storm arrival.

5.7 Caisson Full Weight. Safety Factor against Failure and Additional Consolidation

Consider again in Figure 5.23 the symmetric, shallow, failure mechanism (half width of the caisson). The dissipation work on the edges of the triangular wedges will be approximated by three values (two nodes and the mid point). For instance the average c_u value on segment A′B′ will be

$$c_u^{A'B'} = (c_u^{A'} + c_u^{1} + c_u^{B'})/3, \tag{5.42}$$

where c_u^1 is the strength of the intermediate point between A' and B'. The calculation of the internal dissipation work follows the procedure developed before

$$W_{\text{int}} = \left[\left(c_u^{\text{A'B'}} + c_u^{\text{B'E'}} + c_u^{\text{E'C'}} + c_u^{\text{C'D'}}\right)\right]\frac{b}{4\cos\alpha}\delta_1 + c_u^{\text{B'C'}} b\cos\alpha\delta_1 + c_u^{\text{A'E'}} b\cos\alpha\delta_1 =$$
$$= \left[\left(c_u^{\text{A'B'}} + c_u^{\text{B'E'}} + c_u^{\text{E'C'}} + c_u^{\text{C'D'}}\right)\frac{b}{4\cos\alpha} + \left(c_u^{\text{B'C'}} + c_u^{\text{A'E'}}\right) b\cos\alpha\right]\delta_1. \tag{5.43}$$

The external work, W_{ext}, was given in Equation (5.20). Making $W_{\text{ext}} = W_{\text{int}}$ and isolating the collapse load Q

$$Q = \frac{2}{\sin\alpha}\left(\frac{W_{\text{int}}}{\delta_1}\right). \tag{5.44}$$

The preceding Q value should be minimized with respect to α. This collapse load could be compared with the net caisson stress applied when cells are filled with saturated sand ($Q/b = 220$ kN/m).

An alternative failure mechanism is the non-symmetric zone sketched in Figure 5.24. Mean c_u values on segments limiting the moving triangular wedges are calculated following previous results. The internal dissipation is now given by

$$W_{\text{int}} = \frac{b}{2\cos\alpha}\left(c_u^{\text{AB}} + c_u^{\text{BE}} + c_u^{\text{CD}} + c_u^{\text{EC}}\right)\delta_1 + 2b\cos\alpha\, c_u^{\text{BC}}\delta_1 \tag{5.45}$$

and

$$W_{\text{ext}} = Q\sin\alpha\delta_1. \tag{5.46}$$

Therefore, since $W_{\text{int}} = W_{\text{ext}}$,

$$Q = \frac{1}{\sin\alpha}\left(\frac{W_{\text{int}}}{\delta_1}\right). \tag{5.47}$$

Again, a minimization process with respect to angle α should be performed. The two cases (symmetric and non-symmetric mechanisms) were solved in an Excel sheet following the preceding methodology and equations. The following results are obtained for $t = 14$ days:

Symmetric mechanism

$$Q/b = 264 \text{ kN/m}; \quad SF = \frac{264}{220} = 1.20; \quad \alpha = 56^\circ.$$

Non-symmetric mechanism

$$Q/b = 244 \text{ kN/m}; \quad SF = \frac{244}{220} = 1.11; \quad \alpha = 47^\circ.$$

The variation of the failure load with angle α for the two mechanisms is given in Figure 5.31. The critical failure mechanism is now the non-symmetric one. The reason is that a deeper non-symmetric mechanism goes in search of lower strength values if compared with the shallower symmetric one. This is an effect of the increase in undrained strength, which is maximum at the upper dissipation boundary (where mean stresses are maximum and the drainage more effective) and decreases with depth.

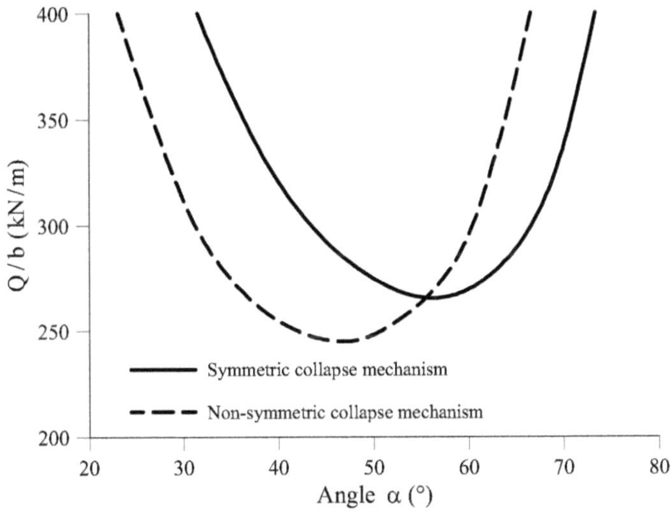

Figure 5.31 Variation of upper bound collapse load with angle α of the failure mechanism. Caisson partially consolidated under full weight.

Though the calculated minimum safety factor is small (≈ 1.1), the fact is that caissons survived also the sand filling and a new consolidation process started under the added load. This result can also be interpreted in the sense that the upper bound calculation developed is actually not that much above the true bearing capacity!

5.7.1 Caissons under full weight

The best way to analyze the consolidation in the days that followed sand filling is to imagine the process divided into two parts. The first one is simply the continuation of the consolidation induced by caisson sinking (caisson filled with water). The second one is a new process whose starting time is the application of the new load. It has been calculated before that filling the caissons with sand implied an increase in uniform load amounting to $\Delta Q/b = 86$ kPa on the base of the caissons.

The previous analysis is repeated having in mind that the objective is now to determine the undrained strength distribution at the time of the storm arrival. Therefore, the distribution of mean stresses will now be given by Equation (5.27) for $q = Q/b = 220$ kN/m. The pore pressures are the result of two contributions:

a) A consolidation process induced by excess pore pressures calculated for $\Delta q = \Delta Q/b = 134$ kN/m, lasting for $t = 21$ days.

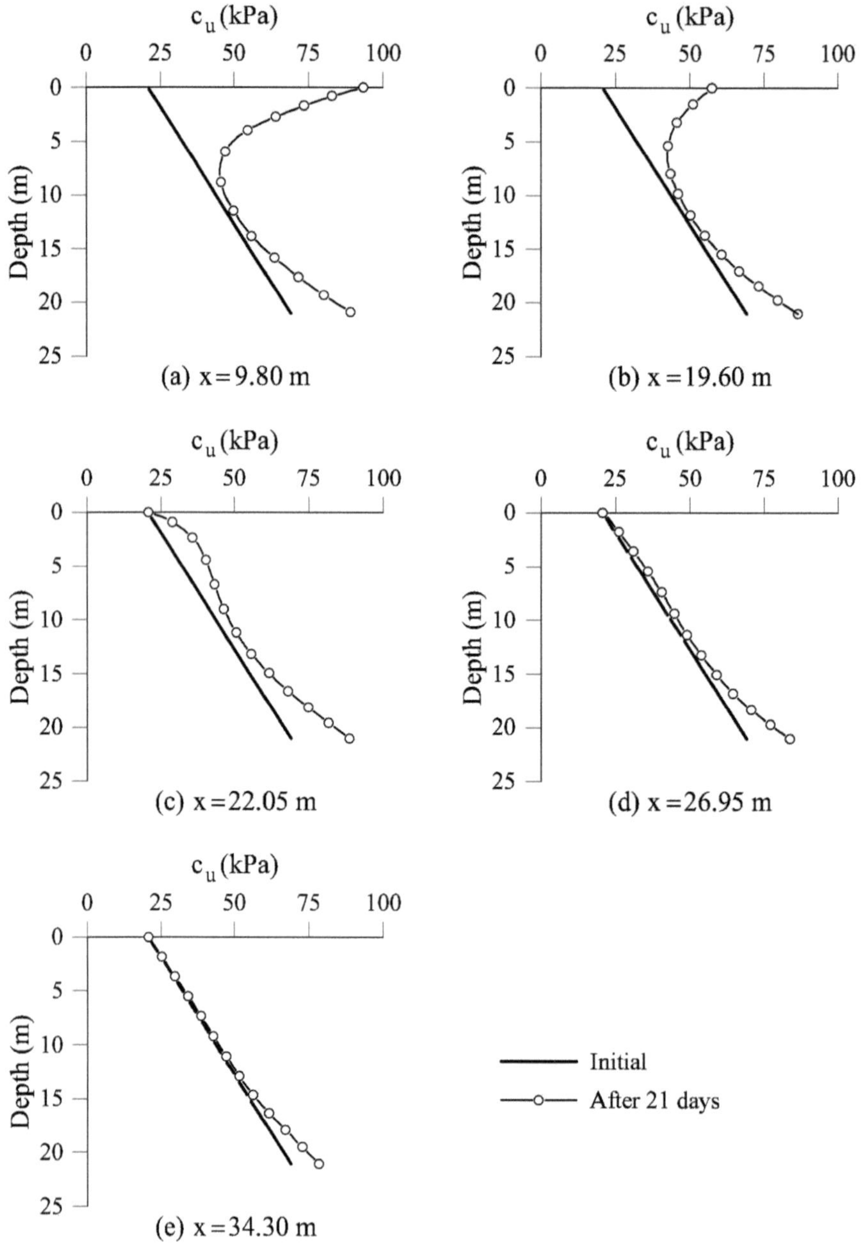

Figure 5.32 Undrained strength at the vertical profiles defined by x = 9.8, 19.6, 22.05, 26.95 and 34.30 m in Figure 5.25 at the time of storm arrival (t = 21 days after caisson sinking). Also indicated is the initial distribution of c_u.

b) A consolidation process induced by excess pore pressures calculated for $\Delta q = Q/b = 86$ kN/m, lasting for $t = 7$ days.

The analysis follows steps previously described and is not repeated here. The resulting distribution of undrained strength is calculated as follows

$$\begin{aligned} c_u(21\text{ days}) &= c_u(\text{initial}) + \\ \Delta c_u(t &= 21\text{ days; water filled caissons}) + \\ \Delta c_u(t &= 7\text{ days; sand filled caissons}). \end{aligned} \tag{5.48}$$

The result of this calculating process is given in Figure 5.32. It shows the resulting current profiles of undrained strength and the initial distribution of c_u for a few vertical profiles characterized by the horizontal coordinates x = 9.8, 19.6, 22.05, 26.95 and 34.30 m (Fig. 5.25). The added load (sand filling) had a limited time to be transformed into effective stresses and, accordingly, into available undrained strength. Nevertheless the upper levels of the foundation soil under the caisson significantly increased the undrained strength. The strength profile shows a maximum at the caisson-foundation contact. Strength decreases rather fast with depth to meet the initial values which increase linearly with depth. This distribution is plotted in a two-dimensional cross-section in Figure 5.33. It shows that caisson consolidation was able to build a "strong" soil nucleus in the upper 7 m although strength decreased continuously with depth. Beyond a depth of approximately 10 m, the initial undrained strength is recovered. A small increase in strength is calculated in the vicinity of the lower drainage layer.

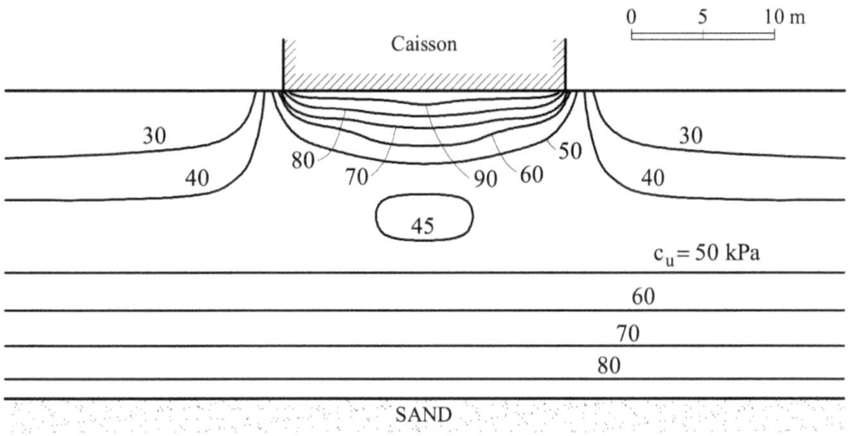

Figure 5.33 Calculated distribution of available undrained strength at the time of storm arrival (November 11, 2001).

The following vertical collapse loads and safety factors were calculated for the strength distribution shown in Figure 5.33,

a) Symmetric mechanism: $q_{\text{collapse}} = Q/b = 313$ kN/m for $\alpha = 58°$,

$$SF = \frac{313}{220} = 1.42;$$

b) Nonsymmetric mechanism: $q_{\text{collapse}} = Q/b = 269 \text{ kN/m}$ for $\alpha = 44°$,

$$SF = \frac{269}{220} = 1.22;$$

This was the situation when the eastern storm of November 11, 2001 hit the caissons.

Note that the critical mechanism is now clearly the deep (nonsymmetric) one. The lower failure surface reaches in this case a depth of 9.46 m, against a maximum depth of 6.72 m for the symmetric, shallower mechanism. This is a consequence of the "inverted" profile of undrained strength created by the caisson weight.

5.8 Caissons Under Storm Loading

5.8.1 Wave forces on caissons

A well-known calculation method to find wave forces on vertical dykes was proposed by Goda (1985). The wave load has two components: an excess pressure on the exposed wall and an uplift pressure on the caisson base. The assumed distribution of these pressures is shown in the sketch of Figure 5.34.

The pressure distributions are equivalent to the concentrated forces:

- Horizontal force H at an elevation Z_H.
- Uplift force U at a horizontal coordinate $X_U = 1/3b$.

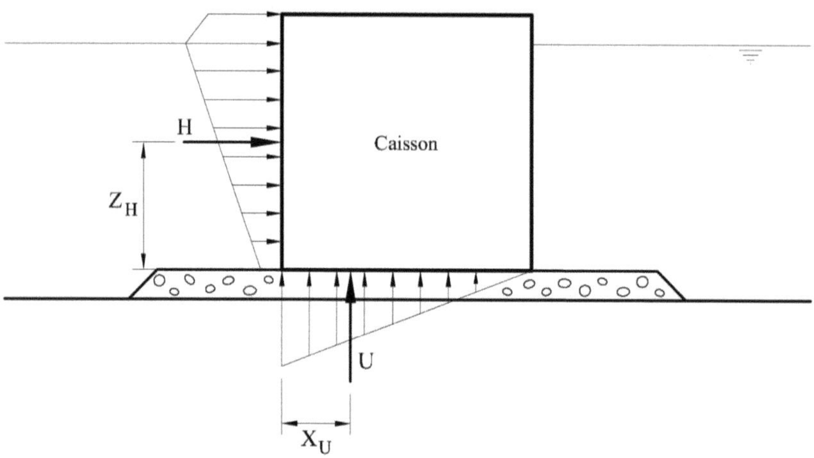

Figure 5.34 Goda's wave pressures.

Goda's formula is detailed in the Appendix to this chapter. For the maximum significant wave height recorded in the storm of November 11, 2001 (see Fig. 5.3), the following forces and points of application are calculated

$H = 779$ kN/m, $Z_H = 8.94$ m,

$U = 267$ kN/m, $X_U = b/3 = 6.5$ m.

The uplift force U reduces the net weight of the caisson. The objective now is to estimate the stability conditions of the caisson under this new set of forces. Wave forces are cyclic, though, and therefore two analyses will be made. First the static stability of the caisson being acted by the highest static forces (H and V) induced by the storm waves will be explored. Then the issue of the soil liquefaction under repeated wave loading will be examined.

5.8.2 Static analysis. Safety factor

The problem is represented in Figure 5.35. The weight Q is now reduced by the calculated uplift force U and a horizontal wave loading H is added. The assumed failure mechanism for the application of the upper bound theorem is also shown. Under the virtual displacement rate δ_1 of the wedge under the caisson, only the vertical force Q_r and the horizontal force H perform external work. Note that in this case, the symmetric mechanism is no longer possible and only the non-symmetric "deep" mechanism will be analyzed. The external work is written

$$W_{\text{ext}} = (Q - U)\delta_1 \sin\alpha + H\delta_1 \cos\alpha. \tag{5.49}$$

The internal dissipation work follows previous developments. Since the soil has a marked non-uniform distribution of undrained strength, the calculations are based on the c_u values estimated at the corners of the mechanism (A, B, C, D, E) and at the mid points (1, 3, 4, 5, 6). Note also that the caisson and the wedge ABE displace as a solid body and therefore no shear dissipation occurs on AE. The dissipation work is therefore given by Equations (5.45) and (5.42).

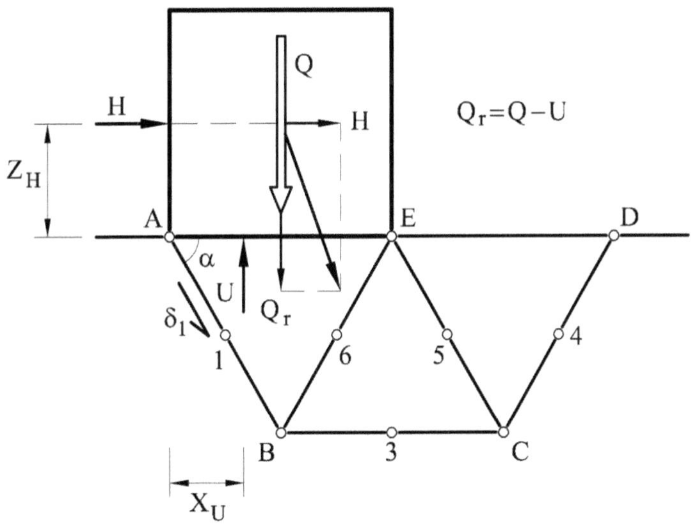

Figure 5.35 Failure mechanism under combined gravitational load and static wave forces.

Now the determination of the safety factor against failure requires more careful consideration. So far in this chapter, the safety factor has been defined as the ratio between the limiting vertical load (determined through the upper bound theorem) and the actual load induced by the caisson defined by its own weight. But now three different forces are acting on the caisson: weight Q, horizontal load H and uplift U.

We will distinguish the real forces from the forces actually inducing the failure (Q^{real} vs Q^{failure}, etc). A single safety factor SF is defined as follows:

$$SF = \frac{Q^{\text{failure}}}{Q^{\text{real}}} = \frac{H^{\text{failure}}}{H^{\text{real}}} = \frac{U^{\text{failure}}}{U^{\text{real}}}. \tag{5.50}$$

The upper bound equilibrium equation ($W_{\text{ext}} = W_{\text{int}}$), which defines the failure loads, may be written as follows:

$$SF\left[\left(Q^{\text{real}} - U^{\text{real}}\right)\sin\alpha + H^{\text{real}}\cos\alpha\right] = \frac{W_{\text{int}}}{\delta_1} = W_{\text{int}}^*. \tag{5.51}$$

This equation defines the safety factor as

$$SF = \frac{W_{\text{int}}^*}{(Q^{\text{real}} - U^{\text{real}})\sin\alpha + H^{\text{real}}\cos\alpha}. \tag{5.52}$$

W_{int}^* is a function of α and the minimization of Equation (5.52) with respect to α will provide the best upper bound approximation to SF.

Consider now the structure of W_{int}^* given in Equation (5.45). It is, in fact, a weighted sum of c_u values taken at different positions (c_{uj}) (the nodes of the mechanism and the auxiliary intermediate points)

$$W_{\text{int}}^* = \sum_{j=1}^{N} \beta_j c_{uj}, \tag{5.53}$$

where β_j are coefficients which depend on the geometry of the mechanism. Therefore, Equation (5.51), which establishes the balance between external work and internal dissipation, can be written

$$SF\left[(Q^{\text{real}} - U^{\text{real}})\sin\alpha + H^{\text{real}}\cos\alpha\right] = \sum \beta_j c_{uj} \tag{5.54}$$

or

$$(Q^{\text{real}} - U^{\text{real}})\sin\alpha + H^{\text{real}}\cos\alpha = \sum \beta_j \frac{c_{uj}}{SF} = \sum \beta_j c_{uj}^*, \tag{5.55}$$

where

$$c_{uj}^{*} = \frac{c_{uj}}{SF} \tag{5.56}$$

is the set of mobilized undrained shear strengths necessary to satisfy the balance implied by the upper bound mechanism in terms of the actual real loads.

But this definition is essentially similar to the safety factor concept used in limit equilibrium analysis, i.e., the reduction factor which has to be applied to the strength parameters in order to achieve strict equilibrium under the actual real loads.

The fact that the safety factor defined in (5.50) as a ratio of loads becomes a strength reduction factor in (5.56) is simply a consequence of the linear relationship between failure loads and undrained strength. This is the case in undrained stability analysis but it is far from being so in drained analysis because failure loads and the ($\tan\phi'$) drained strength parameter are not linearly related.

The expression (5.56) has to be seen as a sound "geotechnical" definition of safety factor. It is equivalent to loading ratios (Eq. (5.50)) only under conditions of undrained analysis.

But the safety factor is not a theoretical or particularly well-defined measure of uncertainty. Other alternatives may suit the needs of the designer. For instance, in the case discussed now, it may be argued that the wave loading is rather uncertain compared with the caisson weight Q^{real}. It may be also accepted that the soil undrained strength is well known. Then, a safety factor aimed at judging the risk of failure induced by wave loading may be defined as

$$SF^{*} = \frac{H^{\text{failure}}}{H^{\text{real}}} = \frac{U^{\text{failure}}}{U^{\text{real}}}. \tag{5.57}$$

Then, following the previous steps, SF^{*} is obtained as

$$SF^{*} = \frac{W_{\text{int}}^{*} - Q^{\text{real}} \sin\alpha}{H^{\text{real}} \cos\alpha - U^{\text{real}} \sin\alpha}. \tag{5.58}$$

This function of α should be now minimized, following the procedures associated with the upper bound theorem.

Safety factors SF and SF^{*} were calculated for the following set of forces: Q^{real} = 4,312 kN/m; H^{real} = 779 kN/m; U^{real} = 267 kN/m and for the calculated distribution of c_u values at the time of the storm arrival (t = 21 days after first sinking).

The calculated Safety Factors and angle α, which defines the mechanism, are

$$SF = 1.10 \quad (\alpha = 41^{\circ}), \quad SF^{*} = 1.77 \quad (\alpha = 41^{\circ}).$$

The calculation indicates that the static wave force was not enough to induce a generalized failure of caissons but it was quite close, if one considers the classical definition of safety factor (Eq. (5.56)). It is worth reminding that our upper bound mechanisms provide estimates that are pretty close to the true ones: after the caisson sand filling we also predicted the SF = 1.1 and the fact was that the caissons did not fail!

The fact that SF^* is a high value indicates that the scale to measure risk should be based on a given definition of safety factor. If the definition changes, we should be prepared to change the scale of risk. In this book Safety Factor is a strength reduction factor, as defined, for undrained conditions, in Equation (5.56).

The deep failure of caissons was also an indication that the foundation soil had experienced an additional reduction in strength most likely associated with soil liquefaction. The next section deals with the liquefaction of the foundation soil.

5.8.3 Analysis of liquefaction

Stability of caissons subjected to horizontal loading is checked in practice by examining a few failure possibilities: bearing capacity (or overall stability), sliding on its base and overturning. In the preceding section the bearing capacity of caissons against its own weight, combined or not with a static estimation of wave loading, has been discussed. The survey after the failure did not provide any support for sliding or overturning modes of failures.

The deep sinking of caissons after the failure (Fig. 5.5) and the type of failure (tilting of caisson top in the seaward direction) suggest that soil liquefaction played a significant role in the failure. The cyclic interaction diagram (Fig. 5.13) shows that liquefaction conditions are defined by the static or average shear stress ratio, the cyclic ratio, and the number of applied cycles.

Instead of performing a comprehensive dynamic analysis, the following simplified approach will be followed here:

a) Shear stresses on horizontal planes (τ_{xz}) will be computed on the foundation for the following two states:

a1) caisson weight;

a2) wave action.

The theory of elasticity will be used in these calculations.

b) Stress ratios (shear stress over vertical effective stress) will then be calculated. The distribution of vertical effective stresses (σ'_z) corresponds to the consolidation time at the time of storm arrival ($t = 21$ days after caisson sinking).

c) The calculated stress ratios will be compared with the information provided by the cyclic interaction diagram (Fig. 5.13). For the estimated number of loading cycles applied by the storm, points in the foundation soil may either fall in a stable zone or in an unstable (liquefied) domain. Points "satisfying" the liquefaction condition will define an area where undrained soil strength will decrease to a post liquefaction state.

d) A new stability analysis following the upper bound methodology will be carried out. The spatial distribution of undrained strength will now be a consequence of the initial state (linear increase of c_u with depth), the previous consolidation history under caissons weight, and the c_u reduced values on the liquefied areas.

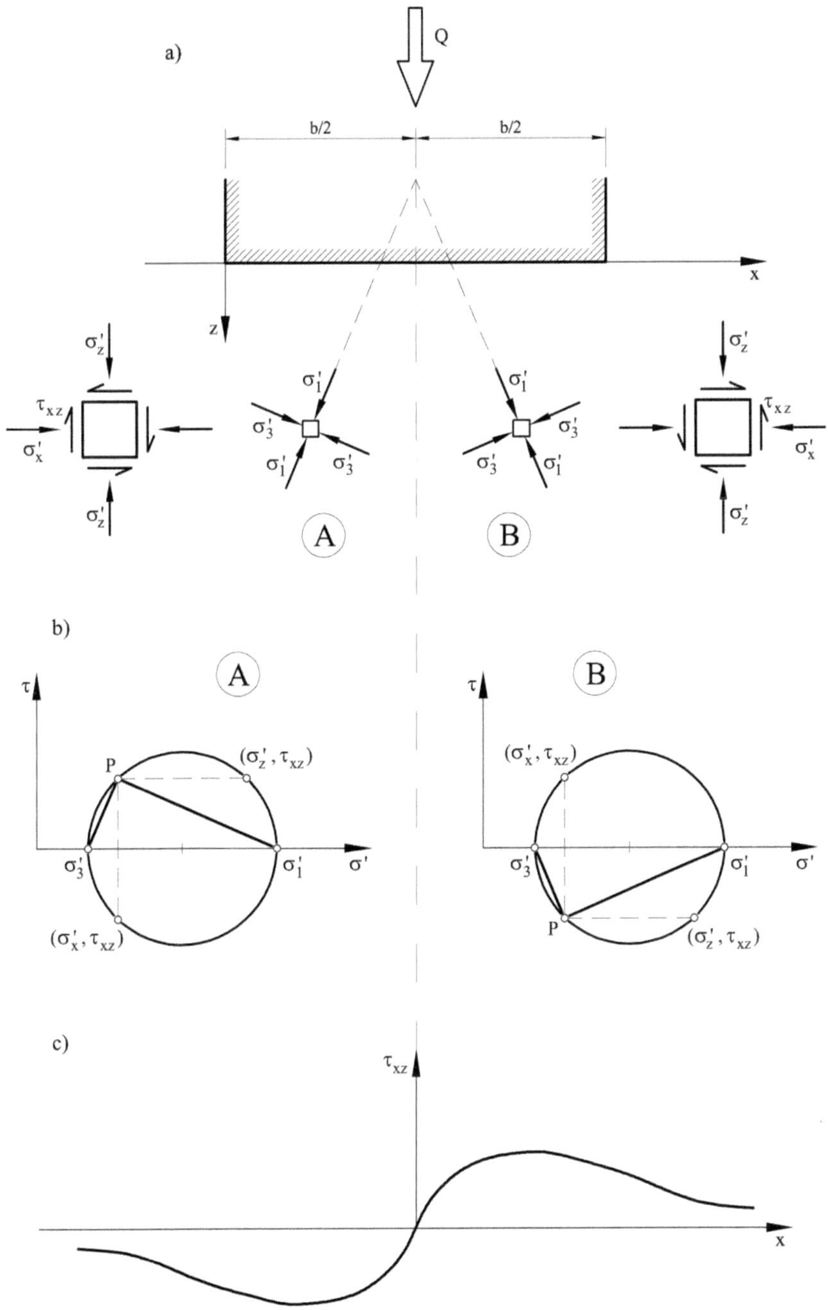

Figure 5.36 (a) Stress distribution under the caisson weight; (b) Mohr diagrams; (c) distribution of shear stresses on a horizontal plane.

a) Shear stresses on horizontal planes

Consider the stress distribution under the caisson induced by its own weight. A descriptive representation of the principal stresses inside the ground is given in the sketches of Figure 5.36. The major principal stress is directed towards the axis of the load, on the surface. Two points are indicated in the figure (A and B) symmetric with respect to the caisson axis. The two Mohr circles on the same figure (Fig. 5.36b) provide the normal and shear stresses on any plane through points A or B. The circles are oriented by means of the pole P, also plotted in the figure.

In point B, the shear stress τ_{xz} on horizontal planes is directed in the direction of the increasing x coordinate. However, in the symmetric point A, τ_{xz} has the opposite sign. On the vertical caisson axis, the major principal stress is vertical and the shear stresses on horizontal or vertical planes are zero.

Therefore, the caisson vertical loading induces a distribution of shear stresses on horizontal planes which is point-symmetric with respect to the origin of the plot (Fig. 5.36c). At the origin the shear stress is zero and at increasing x distance (in both directions, negative and positive), away from the caisson foundation area, they will eventually vanish. Therefore, the shear stress, when plotted along straight lines parallel to the x-coordinate, will start at zero at $x = b/2$ and it will find a maximum at some given distance from the axis to decrease again with distance. The plot in Figure 5.36c is a qualitative representation of τ_{xz} on a horizontal axis that follows the preceding observations.

The distribution of τ_{xz} beneath a uniformly loaded infinite strip was given in Equation (5.26d). The calculated shear stress on several horizontal planes under the caisson weight, immediately before the storm, is given in Figure 5.37. These shear stresses are regarded as the static permanent shearing in the ground. Note that the initial geostatic states, where effective principal stresses follow the vertical and horizontal directions throughout the soil, do not introduce any shearing on horizontal planes.

Note the skew-symmetric distribution of shear stresses, and the existence of a maximum/minimum at some distance from caisson axis, as predicted. Close to the foundation surface, a sharp peak is calculated at the transition between the loaded and the unloaded areas. Away from that zone of intense shearing, the absolute value of the shear stress is small. At depth, the peak "widens" and the shear stresses distributes more evenly. The caisson weight induces significant shear stresses at depth. Even at a depth of 20 m, where the sandy stiff layer marks the lower boundary of the silty foundation soil, the maximum shear stress is close to 40 kPa.

The fact that a shear stress is positive or negative is not of particular significance here because they are equally capable of inducing limiting conditions (isotropic soil properties are assumed).

Consider now the wave action in Figure 5.38. The resultant horizontal force H acting at a height Z_H over the caisson base is made equivalent to a shear force H at the foundation base and a moment $M_H = H \cdot Z_H$. A triangular distribution of water overpressures at the caisson-soil interface is also acting, following Goda (1985). It is equivalent to a uniform uplift load U and a moment M_U, which adds to M_H. The value of these forces for the storm of November 11, 2001 is calculated in the appendix.

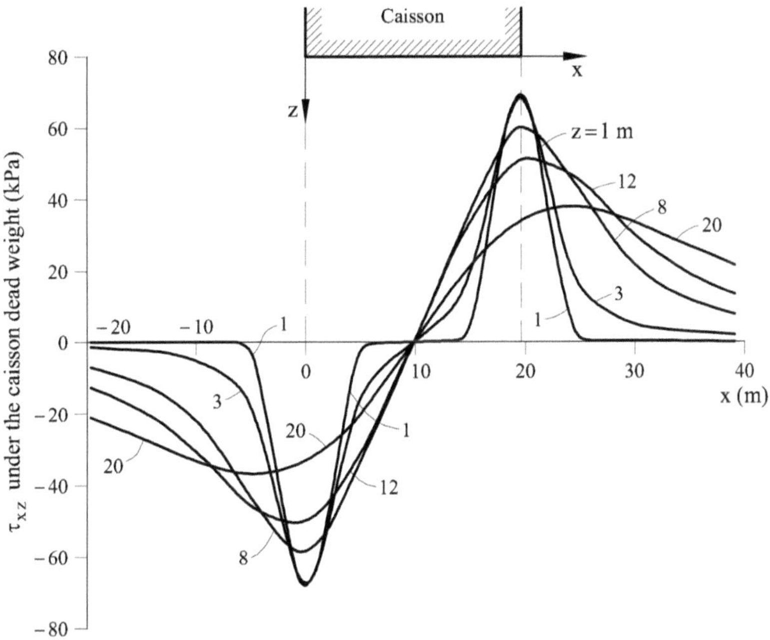

Figure 5.37 Calculated τ_{xz} values under the caisson dead weight at depths z = 1, 3, 8, 12 and 20 m.

The set of wave induced forces is equivalent to the following set of stresses on the caisson foundation interface:

I: A shear stress $q = H/b$ in the direction of H;

II: A uniform uplift stress $p = U/b$;

III: A uniform uplift stress $p_1 = \sigma_{max} = 6M/b^2$, where $M = M_H + M_U$;

IV: A triangular distribution of normal compression stresses with maximum value $p_2 = 2\sigma_{max} = 12M/b^2$.

Stress distributions inside the soil for cases II and III were already given in Equation (5.26). Poulos and Davis (1973) also provide an elastic solution for Cases I and IV. The shear stress τ_{xz} for these two loading cases is given in Figure 5.38c.

Figure 5.39 shows the calculated shear and normal stresses for the following cyclic forces H = 779 kN/m; $M_H = H{\cdot}Z_H$ = 7,242 kN·m/m, U = 267 kN/m; M_U = 869 kN·m/m. These values are taken from the appendix. Now the skew-symmetry of shear stress distribution is lost. However, the peak values for shallow depths concentrate again in the vicinity of the caisson edges. In addition, the shearing force applied at the interface results in significant shear stresses at shallow depths, under the caisson base. Cyclic shear stresses are of the same order of magnitude as static shear stresses.

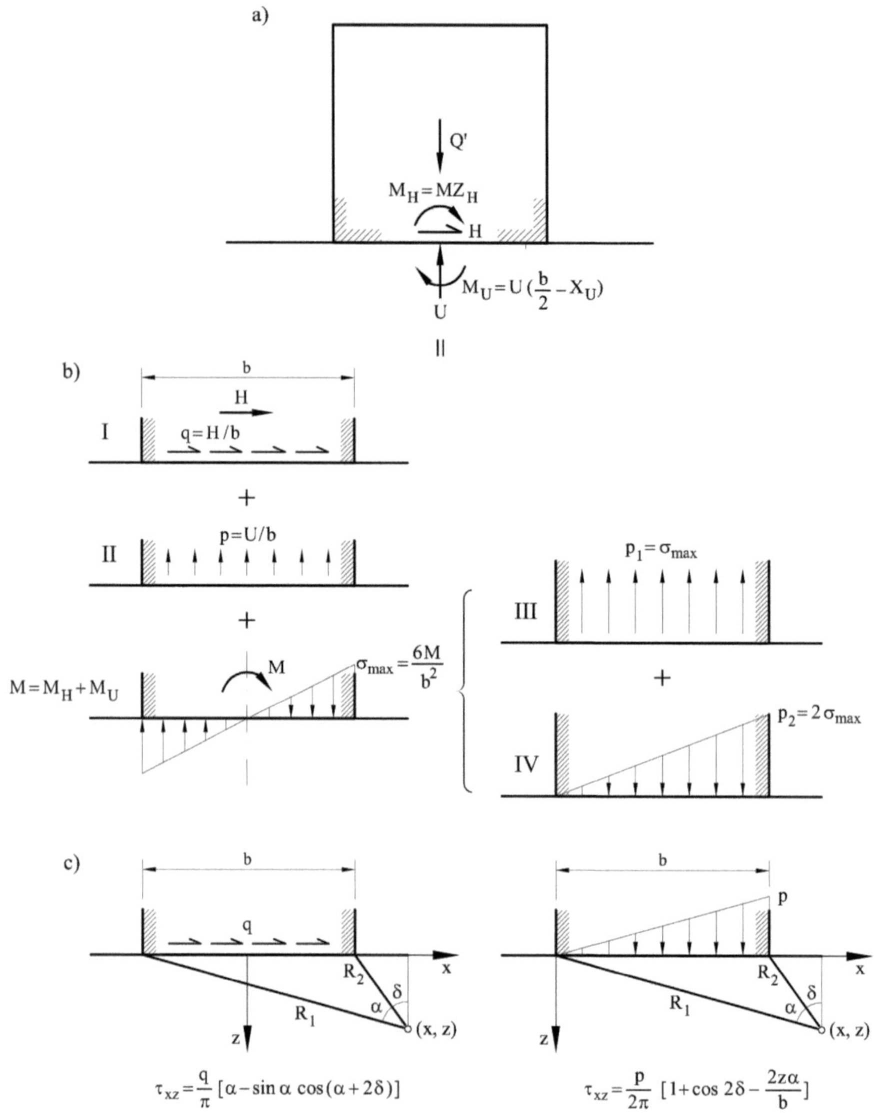

Figure 5.38 (a) Forces in caisson in the presence of wave loading; (b) stresses on the caisson-foundation interface; (c) shear stresses, τ_{xz}, in the subsoil induced by uniform distribution of boundary shear forces and a triangular normal loading.

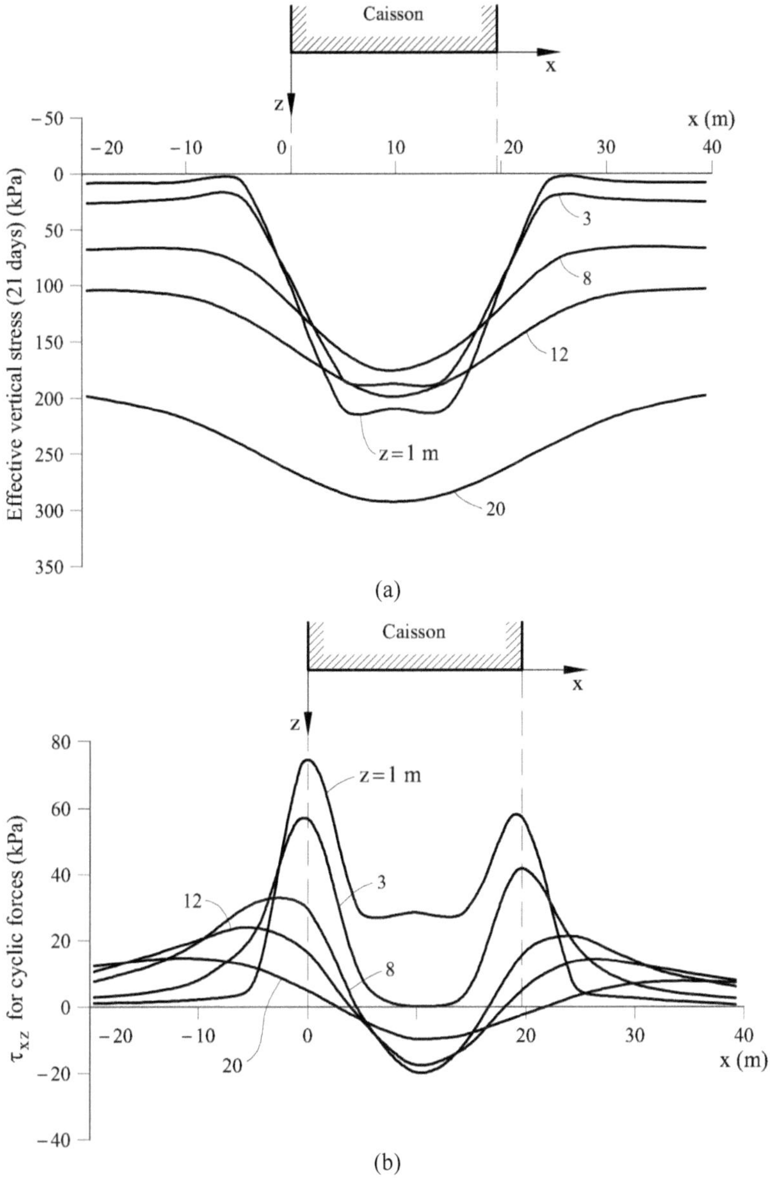

Figure 5.39 Calculated (a) static effective vertical stress σ'_v and (b) cyclic shear stress τ_{xz} at depths z = 1, 3, 8, 12 and 20 m.

b) Stress ratios

Normal effective stresses on horizontal planes were calculated previously. Since they are required to normalize shear stresses, they have been plotted on similar positions in Figure 5.39b. A vertical profile is also given in Figure 5.40 to point out that vertical effective stresses reach a high value close to the caisson base (because of the full consolidation at shallow depths). They decrease with depth down to a depth of about 7 m and increase when the geostatic weight – initial state – dominates again.

The ratios τ_{ave}/σ'_v and τ_{cycl}/σ'_v are now calculated. They are plotted for the same depths in Figure 5.41a,b. The two figures show that stress ratios reach very high values, at shallow depths, in the transition zone between loaded and unloaded areas. Of course, these high stress ratios cannot be resisted by the soil and stress redistributions will occur. The figure indicates, however, that these shallow zones at the edge of caissons are plastified and, in addition, they are particularly critical against cyclic loading. Under the caisson, the stress ratio decreases rapidly due to the high confining stress. At increasing depths the transition between loaded-unloaded surface areas remains in critical conditions. This is better appreciated if the stress ratios in Figure 5.41 are compared with the cyclic interaction diagram in Figure 5.13.

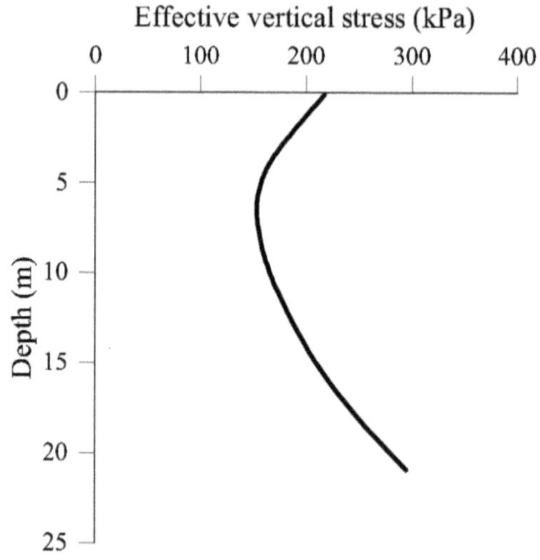

Figure 5.40 Effective vertical stresses at $t = 21$ days after caisson sinking at $x = 14.70$ m.

c) Liquefied zones

Calculated stress ratios were compared with the cyclic interaction diagram given in Figure 5.13. Stress ratios falling on the unstable or liquefied zone have bean indicated in Figure 5.42. The "stable" zone is defined in a simplified manner by the two conditions $\tau_{cycl}/\sigma'_v < 0.15$ and $\tau_{ave}/\sigma'_v < 0.25$. Outside this rectangular domain, the soil is assumed to reach liquefaction. The limiting cyclic stress ratio (0.15) is approximately valid for a number of cyclic stress applications of 100 – 200. This is close to the number of wave impacts when the storm reaches its maximum intensity (a significant wave height of 4 m).

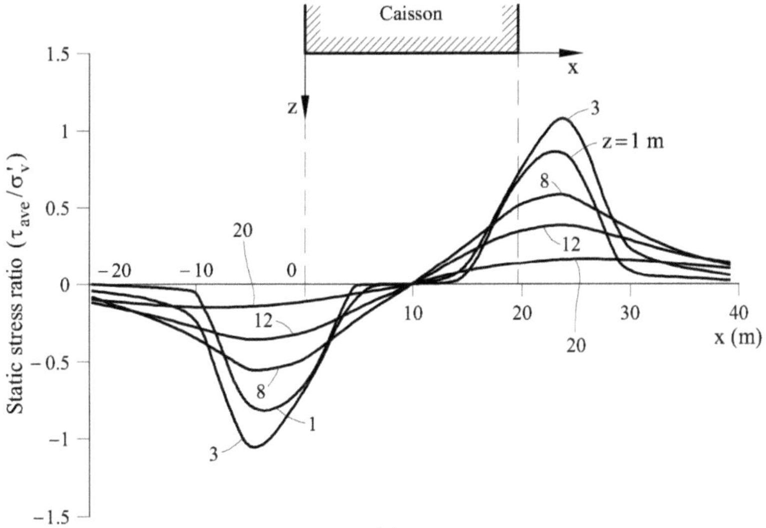

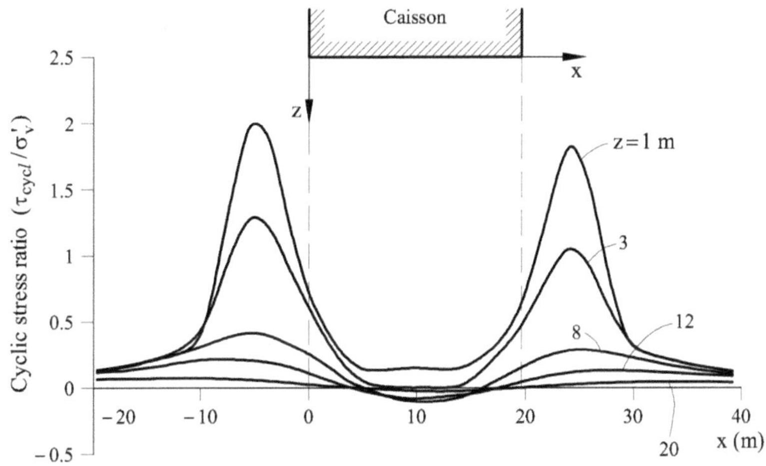

Figure 5.41 (a) Static (or average) stress ratio; (b) cyclic stress ratio.

The liquefied areas define two wide bulbs on both edges of the caisson. They reach a maximum depth of about 14 m. Under the caisson central area, however, no liquefaction conditions develop (Fig. 5.42).

The undrained strength of the liquefied areas was given in Figure 5.14. The scatter is high and it is difficult to select a value with some confidence, but a post liquefaction strength close to the maximum suggested by this plot $c_u = 0.06\sigma'_v = 0.09\sigma'_m$ was adopted to estimate the stability in the case of soil liquefaction.

d) Stability of caissons

The final step of the analysis is to calculate the failure conditions when part of the foundation soil is liquefied. The failure mechanism crosses the liquefied zone located on the leeward side of the caisson foundation (Fig. 5.42). In the non liquefied areas, the distribution of undrained strength was reported before. Within the liquefied area $c_u = 0.09\sigma'_m$.

The calculated safety factor, defined as in Equation (5.52), drops to 0.56 for a critical mechanism defined by an angle α = 44°. Even if the liquefaction zone is characterized by a relatively high value of the residual undrained strength ($0.09\sigma'_m$), the drop in safety factor is very significant.

If a more substantial reduction of strength is specified, the safety factor drops to very low values. This result helps to explain the deep burial of caissons as a consequence of the failure.

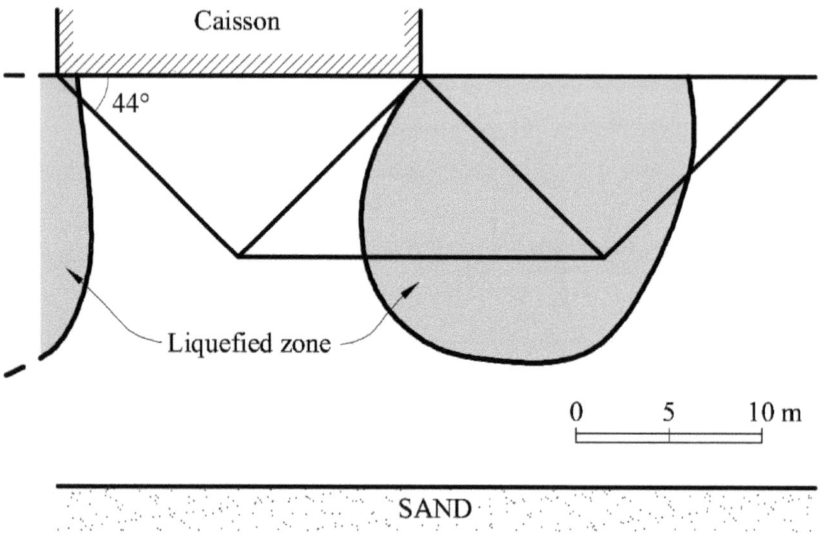

Figure 5.42 Liquefied areas under a caisson induced by storm cyclic loading. Also indicated is the critical failure mechanism.

5.9 Discussion

The upper bound theorem of plasticity provides an unsafe estimate of bearing capacity. A comparison against the available analytical solution for increasing undrained soil strength shows that errors may be in the order of 17% for the failure mechanism selected here. The mechanism is, however, simple and it may accommodate, at a limited effort, complicated spatial distributions of undrained shear strength. Consider, however, this limitation from a different perspective. Figure 5.43 shows the variation in time of the safety factor of the caissons, starting at the time of caisson sinking (t = 0) and ending at the failure time, 21 days later.

Symmetric (shallower) and non-symmetric (deeper) mechanisms are considered when appropriate. The plotted variation of safety factor with time follows the minimum values

The calculated safety factors, taking the upper bound solution as a correct estimate, should be regarded in sequence, paying attention to the changes in *SF* but not to its precise values. Its evolution in time is the most valuable information. The fact that caissons did not fail during the first sinking provides a reference condition ($SF > 1$) for $t = 0$, irrespective of calculations. Presumably, in view of the results obtained, the actual value of *SF* was in the vicinity of 1.3. Then the calculation process described, which maintains the failure mechanism, results in a *SF* value, at $t = 21$ days and for a static equivalent loading to wave action, which is greater but close to one.

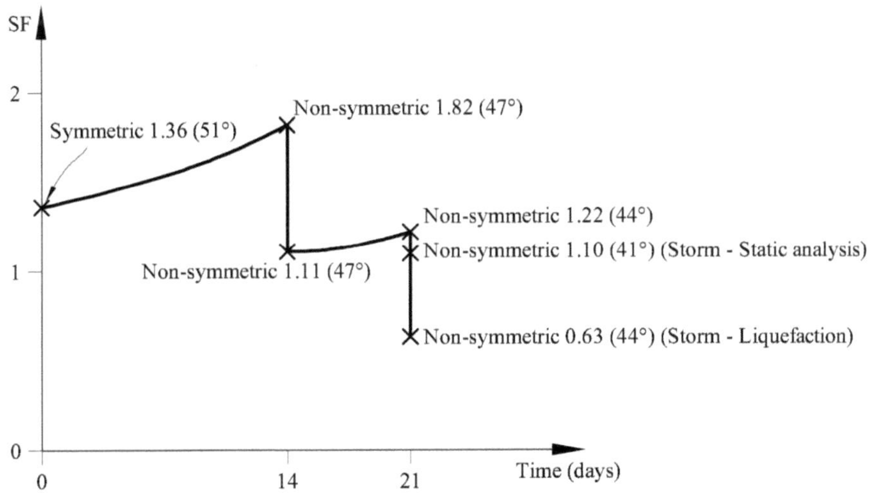

Figure 5.43 Evolution of caisson safety factor (*SF*) against bearing capacity failure. The value in parenthesis indicates the critical α angle of the failure mechanism.

Therefore, it is concluded that the failure and the deep sinking of caissons should be explained by a strength-loss mechanism associated with the cyclic wave loading. A soil liquefaction phenomenon seems a good candidate to explain the strength loss, especially if one considers the deep burial of failed caissons into the foundation soil. Of course, the silty deltaic deposits seem prone to liquefy in view of the accumulated experience synthesized in Figure 5.13. The consideration of liquefaction, which was analyzed by a procedure inspired in the cyclic mobility diagram of the natural soil, indicates that extensive zones of the caisson foundation soils reached critical conditions under the storm of November 11, 2001. A substantial reduction in safety factor is calculated, even if the loss of undrained strength is moderate. In other words, irrespective of the safety factor at $t = 21$ days under a pseudostatic wave loading, the liquefaction of the foundation soil is capable of inducing a catastrophic failure. These comments help to accept the upper bound calculation as a useful and simple procedure to analyze the conditions and reasons for the failure, even if some error is experienced in the actual estimation of failure load.

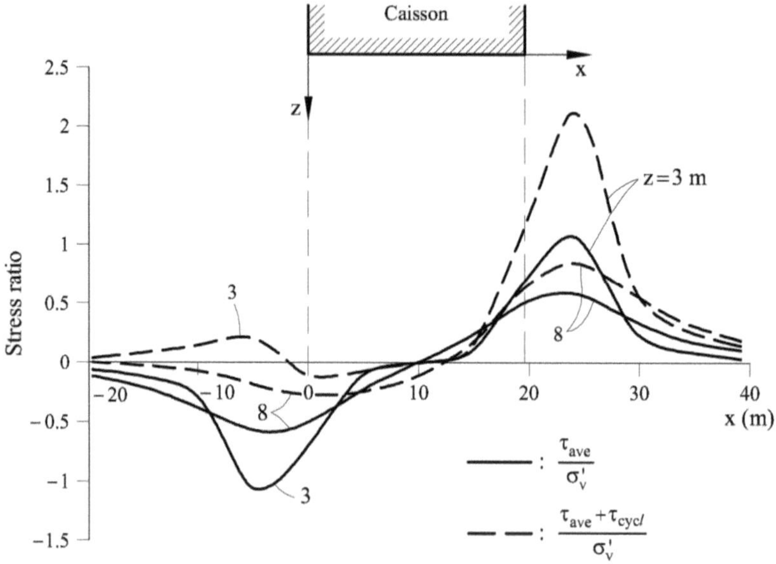

Figure 5.44 Distribution of average and maximum stress ratios.

Regarding Figure 5.43 some uncertainties remain on the exact timing of sand filling. But this is not particularly relevant because the caissons did not fail under the full weight. If a shorter time elapsed between water filling and sand filling and this shortened interval had led to $SF < 1$ at the time of sand filling operation, it simply implies that the initial SF was somewhat higher than the value calculated here. The procedure developed can be applied to a different sequence of events.

The kinematic mechanisms analyzed in this chapter imply a vertical displacement of caissons. The actual failure mechanism involved also a rotational component (see Fig. 5.5) perhaps associated with soil liquefaction. The mechanisms selected are only an approximation but they are capable of explaining in a satisfactory manner the sequence of events which led to the failure.

The liquefaction analysis has also limitations. The large stress ratios calculated in some localized areas of the foundation (in the vicinity of the caisson's edges) cannot possibly occur because soil yielding would redistribute the stress field. But classical elastic stress distributions (which are independent of elastic constants for homogeneous soil profiles) provide a reasonable approximation for extended areas of the foundation soils.

On the other hand, the plots of average static stress ratios and of the maximum stress ratios (average + cyclic) in Figure 5.44 indicate that, in some zones of the foundations soil, shear stress reversals did occur. In fact, it has been found that shear stress reversals (changing sign of τ_{xz}) lead to a more efficient and rapid development of liquefaction conditions (see Section 5.12 on Advanced Topics below). The testing program leading to the cyclic interaction diagram of Figure 5.13 did not consider shear stress reversals and this may be also a source of error. In this case, actual field conditions seem somewhat more critical than experimental ones.

The "signature" of wave impact loading is known to be far from a regular sinusoidal or regular variation of stress, as imposed in cyclic simple shear tests. Figure 5.45 shows the time development of impact forces for a storm of a significant wave height of 11 m (Alonso and Gens, 1999). These are different conditions but the plot indicates that field impacts tend to concentrate on short time intervals. In the remaining time, within the wave period, wave forces remain small. This is also a difference between laboratory testing and field conditions.

The four failed caissons were essentially isolated and surrounded by open sea conditions. Therefore, only wave forces in the landward direction should be considered. If caissons protect a water body, the wave trough results in an unbalanced hydrostatic load in the seaward direction. This load, which enhances stress reversals, was unlikely in this case.

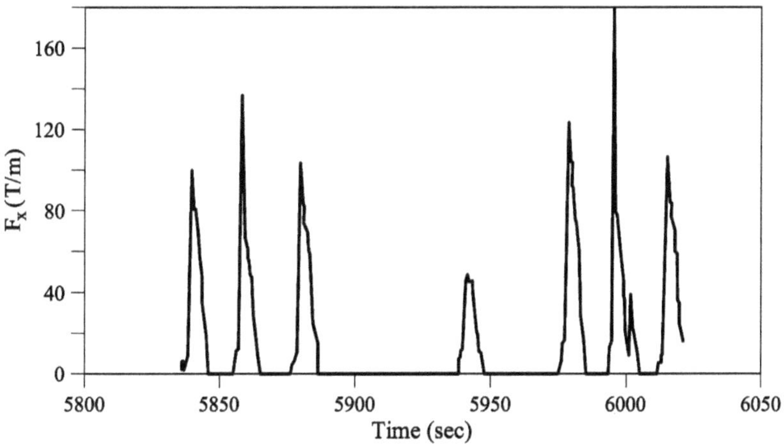

Figure 5.45 Partial record of impact wave forces against a dyke (Alonso and Gens, 1999 © 1999 Taylor and Francis Group. Used with permission).

5.10 Mitigation Measures

The failure could have been avoided by one or more of the following mitigation measures:

5.10.1 Increasing the consolidation time under caisson weight

Then the undrained strength of the natural soil would increase as well as the safety factor. However, if nothing else is changed in the design, the risk of failure remains high because the time of storm arrival is not under control. If caissons are sunk in place at, say, the beginning of summer time in the Mediterranean environment of Barcelona, chances are that no strong storms will hit the works for a few months. Then a high degree of consolidation will be achieved. It remains to be checked if a design storm is resisted in this case. But, most probably, it will not be the case if liquefaction develops.

5.10.2 Increasing the size of the granular berm

Both its thickness and its lateral extension should be properly designed. Increasing the thickness adds a drained frictional resistance to any potential failure surface crossing the granular layer. Increasing the lateral extent increases the size of the potential failure surface or forces it to cross the granular berm. This is a good solution but there are a few remarks which should be taken into account:

- Increasing berm thickness without previous dredging may impair navigation requirements around caissons.
- Dredging soft, normally consolidated soils, requires very gentle slopes (around 1:10) to ensure stability. Environmental issues (water contamination) should be addressed.
- If dredged materials are substituted by a granular layer, the undrained strength of the underlying natural soil is essentially unchanged because effective stresses will not change much (unit weights of granular soil and a soft clayey soil are similar).
- Consolidation times are reduced significantly if vertical drains at close spacing (say 1 – 3 m) are installed in advance to any overloading.

5.10.3 Improving foundation soils

This is an expensive option. Preloading would increase c_u but this is a lengthy and costly operation in a sea environment. It has associated drawbacks (impairing navigation, increasing turbidity of waters). Soil strength may also be improved by other techniques: installing piles or granular columns or performing a dynamic consolidation by pounding the soil.

5.10.4 Increasing caisson width

Increasing the caisson width contributes to increase in the bearing capacity and it also reduces the risk of liquefaction. But the improvement is relatively small and significant increases in caisson width will be required to guarantee adequate safety.

5.10.5 After the failure

Caissons were demolished "in situ" by repeated hammering by a falling dead weight operated from a floating barge. Then they were covered by a random rockfill. An embankment-type of breakwater was finally built on the position of Caissons 1 to 4. The buried part of failed caissons remained in place.

5.11 Lessons Learned

5.11.1 Foundation on normally consolidated soft soil

The design of safe foundations for caissons dykes in an open sea environment sitting on normally consolidated soft silty and silty clay deposits of low plasticity poses a significant geotechnical challenge. The case described illustrates the difficulties for a proper evaluation of the safety factor against global (or bearing capacity) failure under caisson self-weight and wave loading.

5.11.2 Strength changes due to caisson loading

The following aspects should be accounted for in an analysis of global or bearing capacity type of failure:

- The protocol of caisson sinking and subsequent weight increase.
- The consolidation process under the history of caisson increase in weight.
- The strength increase induced by consolidation.
- The forces induced by wave action under the design storm.
- The risk of soil liquefaction under repeated loading (if foundation soils are prone to liquefaction; see Fig. 5.10).

A step-by-step procedure to consider this sequence of events has been described in this chapter. The analysis of the consolidation process has required the development of a closed form solution for the one-dimensional consolidation equations under a general parabolic initial excess pore pressure.

5.11.3 Undrained vs drained analysis

In low permeability soft soils, the most critical limiting condition is an undrained failure. In this case, the external failure loads are linearly related to the undrained strength. This implies that safety factors, calculated as the ratio of failure and applied loads or as the necessary reduction coefficient of the field undrained strength in order to reach strict equilibrium under the actual external loads should be identical. This is not the case if drained strengths are involved (for instance, if a granular embankment is significantly affected by the critical failure).

5.11.4 Evolution of undrained strength

The initial profile of undrained strength of a normally consolidated deep layer of soil (a linear increase of c_u with depth) is substantially modified during the caisson-induced consolidation process. If upward drainage is allowed (a common case), the undrained strength will initially decrease with depth and eventually it will increase again, beyond a certain depth, when the geostatic stress distribution dominates again the profile. Bearing capacity analyses should be able to include these variations for a proper estimation of the evolution of safety with time. Limit theorems of plasticity offer this possibility at a limited calculation effort.

5.11.5 Spatial distribution of c_u controls mode of failure

The critical failure mechanism is controlled by the spatial distribution of c_u. A linear increase of c_u with depth results in shallower mechanisms. A reduction of strength with depth calls for deeper failure mechanisms.

5.11.6 Type of loading and the failure mechanism

Under pure vertical load (self-weight) and a linearly increasing c_u with depth, a double symmetric mechanism with respect to the axis of loading is the most critical one. However, if horizontal loads are introduced, the double symmetric mechanism in no longer possible and a non-symmetric failure mechanism has to be considered.

5.11.7 Alternative definitions of safety factor

Safety factors may be defined in alternative ways. Whenever possible, the classical definition used in limit equilibrium analysis (ratio of field to mobilized strength) should be preferred. The use of partial safety factors (for drained cohesion and friction coefficient) falls into this category. However, the nature of some loading conditions or, indeed, of the failure mechanisms (a relevant example is the failure mode by overturning of walls, which essentially does not involve the foundation soil strength) may require a different formulation. A natural choice is to compare failure loads and actual loads. One example has been given in the section when examining the caisson safety subjected to wave loading and caisson weight. Two warnings may be issued:

- Alternative definitions of safety factor are possible when comparing failure loads and actual loads. Different numerical values of the safety factor reflect the choice made, although the risk of failure should remain unique.
- Safety factors based on strength reduction (as in limit analysis) or in a comparison of loads, are widely different in drained analysis (because of the highly nonlinear relation between failure loads and drained strength parameters) although the risk of failure remains unique.

5.11.8 Defining soil liquefaction conditions

Liquefaction conditions were defined by means of a cyclic interaction diagram. It combines, in a purely experimental criterion, the static, maintained, or averaged stress ratio, the cyclic stress ratio and the number of applied stress cycles which marks the onset of liquefaction conditions. Such a diagram was available for the deltaic silty foundation soils of the caissons.

5.11.9 Simplified analysis of liquefaction

The cyclic interaction diagram has inspired a simplified procedure to analyze liquefaction under field conditions. Shear stresses in the foundation soil were calculated for the permanent and cyclic caisson loading through available closed form solutions in elasticity. The normalizing effective vertical stress was derived from the consolidation analysis. Once the stress ratios are calculated under the caisson, the cyclic interaction diagram provides the liquefied or non-liquefied areas. It was found that liquefaction takes place on both sides of the caisson, under and beyond caisson edges, and at depths that reach the caisson width.

5.11.10 The flexibility of upper bound calculation

The flexibility and capabilities of the upper bound procedure were demonstrated when the liquefied zones were integrated into a failure analysis of the caisson once liquefaction had occurred. A substantial reduction of failure load was calculated for a reasonable estimation of post-liquefaction undrained strength.

5.11.11 Failure mechanism

After the failure caissons were found tilted, their cap displaced in the seaward direction. The failure mechanism implies that both the caisson weight and the wave forces performed a positive work, balanced by the internal dissipation in the soil. No evidence of overturning induced by wave forces was found.

5.12 Advanced Topics

Vertical caisson failures were collected and discussed by Oumeraci (1994). The author describes the history of design and construction of vertical breakwaters and identifies the causes of failures. However, the field information in most cases is very limited especially as far as soil conditions are concerned. He concludes that forces induced by irregular breaking waves are of special concern and are probably a main reason for the observed failures. Concerning the role of the foundation soil, he observes that scour and erosion at the toe of the structures has often been observed. But major failures are probably associated with pore pressure build-up during cyclic loading and eventually with a full or partial liquefaction of the foundation soils. Unlike earthquake loading, wave action is characterized by long time periods (low frequency of loading), a relatively large number of loading applications, and by a sequence of fast wave impacts if a broken wave hits the wall. Once the soil is liquefied or maintains a certain level of pore water overpressure, failure may occur by a critical slip failure mode.

It has also been observed that the direct wave action on the sea bottom may induce the accumulation of excess pore pressures, a phenomenon which may lead to soil instability. Damage of some marine structures has been attributed to this phenomenon which has been reviewed by Jeng (1998, 2001) and Jeng and Lin (2000). Most of the studies refer to sand beds of high porosity. Excess pore pressures eventually dissipate after the wave action and the soil will consolidate, increasing the resistance to a new significant storm.

The irregular wave pattern originated in the vicinity of a vertical wall has also been associated with observed damage. Not only the cyclic loading but the stress rotation experienced by the soil may induce liquefaction (Sassa and Sekiguchi, 2001).

However, the stress changes induced by the wave loading on the marine structures are significantly larger than the (direct) wave-induced loading on the sea bottom (de Groot *et al.*, 2006). This was also the conclusion of large-scale laboratory experiments reported by Kudella *et al.* (2006).

The phenomenon of soil liquefaction has received much attention in geotechnical literature. Most of the experimental information is driven by earthquake research. A comprehensive description of liquefaction behaviour was provided by Ishihara (1993) and Youd and Idriss (2001).

Model testing has provided further insight into the mechanisms of caisson behaviour under wave loading. Scaled model testing under normal gravitational conditions cannot reproduce the stress conditions prevailing at a prototype scale under the foundation and they can hardly reproduce liquefaction behaviour. The type of failure observed in these tests (seaward motion of caisson top) is not supported by some field observations, including the failure reported here.

Centrifuge testing is a powerful tool to study caisson performance under repeated wave loading. Tests have reported by Rowe and Craig (1976), Van der Poel and de Groot (1998) and by Zhang *et al.* (2009). They all report the behaviour of caissons founded on sand beds. Mechanisms of caisson failure by tilting towards the sea direction have been found. Softening and erosion of the sand in the vicinity of the caisson heel, as well as large liquefied zones beneath the caisson edge, have been proposed by these authors. These studies as well as laboratory cyclic shear test on sands, indicate that reversal and irregular shear loading enhances soil liquefaction (if compared with non-reversal regular cyclic loading).

Appendix 5.1 Hydrodynamic Loads on Caisson

The model proposed by Goda (1985) allows transforming wave heights into loads on the structure once the wave height (H_s), wave period (T), and the wave direction are known. In the analyzed case, $H_s = 4.5$ m, $T = 9$ s and the incident angle of the wave with respect to a line perpendicular to the structure, ϑ, is essentially zero. Figure A5.1a shows the pressure distribution according to Goda (1985). The wave load has two components: a trapezoidal excess pressure distribution on the exposed wall defined by the values P_1, P_2, P_3 and P_4, the run up height, η^*, and geometric parameters, and an uplift triangular pressure distribution on the caisson base defined by the value of P_0. The values of the geometric parameters are also indicated in the figure.

The Goda's formulae are written as

$$\eta^* = 0.75(1+\cos\beta)H_D,$$

$$P_1 = 0.5(1+\cos\beta)\left[\alpha_1 + \alpha_2\cos^2\beta\right]\gamma_w H_D,$$

$$P_2 = \frac{P_1}{\cosh\left(\frac{2\pi h}{L}\right)},$$

$$P_3 = \alpha_1 P_1,$$

$$P_4 = P_1\left(1-\frac{h_c}{\eta^*}\right),$$

$$P_0 = 0.5(1+\cos\beta)\alpha_1\alpha_3\gamma_w H_D,$$

with the coefficients expressed as

$$\alpha_1 = 0.6 + \frac{1}{2}\left[\frac{4\pi h/L}{\sinh(4\pi h/L)}\right]^2,$$

$$\alpha_2 = \min\left\{\frac{h_b - d}{3h_b}\frac{H_D^2}{d^2}; \frac{2d}{H_D}\right\},$$

$$\alpha_3 = 1 - \frac{h'}{h}\left(1 - \frac{1}{\cosh\left(\frac{2\pi h}{L}\right)}\right).$$

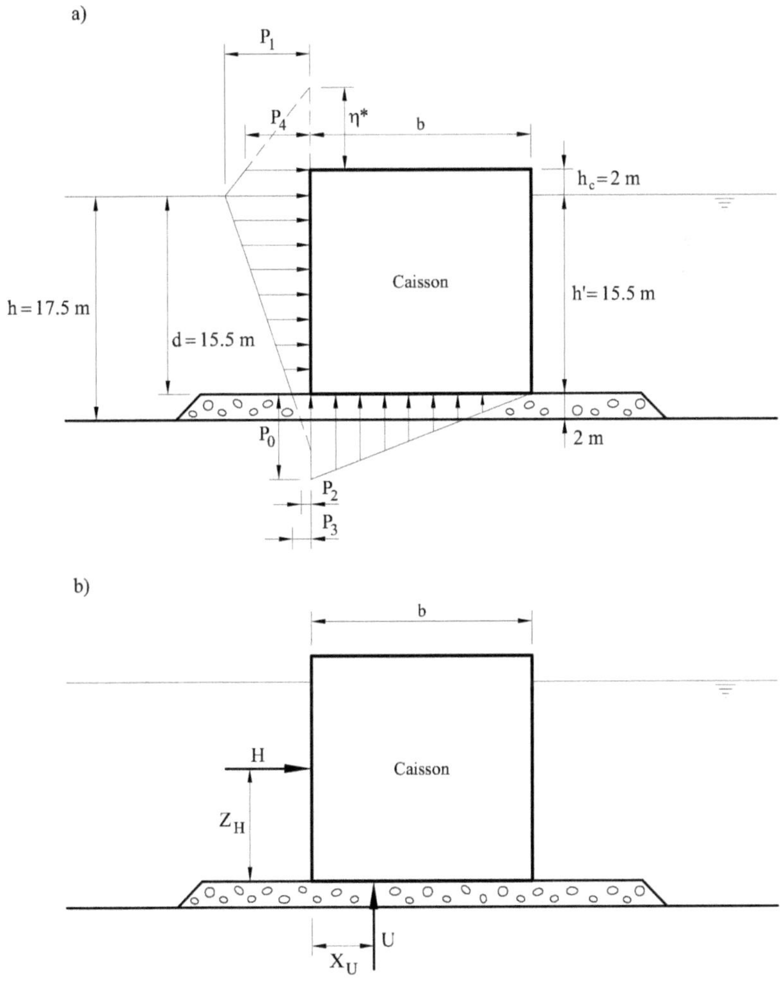

Figure A5.1 (a) Goda's pressure distribution and (b) equivalent concentrated forces.

γ_w is the specific weight of the water (= 10 kN/m^3). The required basic parameters of swell and their values are

$$\beta = \vartheta + 15° = 15°,$$

$$h_b = h + 5 \cdot H_{1/3} \tan\theta = 17.5 + 0 = 17.5 \text{ m},$$

$$L_0 = \frac{gT^2}{2\pi} = \frac{9.8 \cdot 9^2}{2\pi} = 126.34 \text{ m},$$

$$H_{1/250} = 1.8 H_{1/3} \approx 1.8 H_s,$$

$$H_b = 0.17 L_0 \left[1 - \exp\left(-1.5 \frac{\pi h_b}{L_0} \left(1 + 15 (\tan\theta)^{2/3} \right) \right) \right] =$$

$$= 0.17 \cdot 126.34 \left[1 - \exp\left(-1.5 \frac{\pi \cdot 17.5}{126.34} \right) \right] = 10.30,$$

$$H_D = \min\left\{ H_{1/250}; H_b \right\} = \min\left\{ 1.8 \cdot 4.5; 10.30 \right\} = 8.1,$$

$$L = \frac{gT^2}{2\pi} \tanh\left(\frac{2\pi}{L} h \right) = \frac{9.8 \cdot 9^2}{2\pi} \tanh\left(\frac{2\pi}{L} 17.5 \right) \quad \rightarrow \quad L = 78.19 \text{ m},$$

with these values, $\alpha_1 = 0.657$, $\alpha_2 = 0.01$ and $\alpha_3 = 0.524$ and $P_1 = 53$ kN/m, $P_2 = 24.5$ kN/m, $P_3 = 34.8$ kN/m, $P_4 = 44.2$ kN/m, and $P_0 = 27.4$ kN/m.

Once the pore pressure distribution is calculated, the equivalent forces and their position (Fig. A5.1b) can be calculated as

$$H = \frac{1}{2}(P_1 + P_3) h' + \frac{1}{2}(P_1 + P_4) h_c = 778.7 \text{ kN/m},$$

$$Z_H = 9.3 \text{ m},$$

$$U = P_0 \frac{b}{2} = 267.4 \text{ kN/m},$$

$$X_U = \frac{b}{3} = 6.5 \text{ m}.$$

References

Alonso, E.E. and Gens, A. (1999) Geotechnical design and construction of breakwaters. Bilbao Harbour (Invited Lecture). *Proceedings of the 12th European Conference on Soil Mechanics and Geotechnical Engineering.* Amsterdam, 1, 489 – 510.

Alonso, E.E. and Krizek, R.J. (1975) Consolidation of randomly heterogeneous clay strata. *Transportation Research Record* 548, 30 – 47.

Davis, E.H. and Booker, J.R. (1973) The effect of increasing strength with depth on the bearing capacity of clays. *Géotechnique* 23 (4), 551 – 563.

Davis, E.H. and Poulos, H.G. (1972) Rate of settlement under two- and three-dimensional conditions. *Géotechnique* 22 (1), 95 – 114.

Goda, Y. (1985) *Random Seas and Design of Maritime Structures*. University of Tokyo Press, Tokyo.

de Groot, M.B., Bolton, M.D., Foray, P., Meijers, P., Palmer, A.C., Sandven, R., Sawicki, A. and The, T.C. (2006). Physics of liquefaction phenomena around marine structures. *Journal of Waterway, Port, Coastal, and Ocean Engineering* 132 (4), 227 – 243.

Ishihara, K. (1993) Liquefaction and flow failure during earthquakes. *Géotechnique* 43 (3), 351 – 415.

Jeng, D.S. (1998) Wave-induced seabed response in a crossanisotropic seabed in front of a breakwater: an analytical solution. *Ocean Engineering* 25 (1), 49 – 67.

Jeng, D.S. (2001) Mechanism of the wave-induced seabed instability in the vicinity of a breakwater: a review. *Ocean Engineering* 28, 537 – 570.

Jeng, D.S. and Lin, Y.S. (2000) Poroelastic analysis of the wave-sea interaction problem. *Computers and Geotechnics* 26, 43 – 64.

Kudella, M., Oumeraci, H., de Groot, M.B. and Meijers, P. (2006) Large-scale Experiments on pore pressure generation underneath a caisson breakwater. *Journal of Waterway, Port, Coastal, and Ocean Engineering* 132 (4), 310 – 324.

Lunne, T., Robertson, P.K. and Powell, J.J.M. (1997) *Cone Penetration Testing in Geotechnical Practice*. Blackie Academic/Chapman and Hall, E&FN Spon, London.

NGI (2002) *Report on DSS Tests*. Port Authority. Barcelona.

Olson, S.M. and Stark, T.D. (2002) Liquefied strength ratio from liquefaction flow failure case histories. *Canadian Geotechnical Journal* 39, 629 – 647.

Oumeraci, H. (1994). Review and analysis of vertical breakwater failures-lessons learned. *Coastal Engineering* 22, 3 – 29.

Potts, D. and Zdravkovic, L. (1999) *Finite Element Analysis in Geotechnical Engineering: Volume I, Theory*. Telford Publishing, London.

Poulos, H.G. and Davis, E.H. (1973) *Elastic Solutions for Soil and Rock Mechanics*. Wiley. New York.

Rowe, P.W. and Craig, W.H. (1976) Studies of offshore caissons founded on Ostercheelde sand. *Design and construction of offshore structures*, ICE, London, 49 – 60.

Sassa, S. and Sekiguchi, H. (2001). Analysis of wave-induced liquefaction of sand beds. *Géotechnique* 51 (2), 115 – 126

Seed, R.B., Cetin, K.O., Moss, R.E.S., Kammerer, A.M., Wu, J., Pestana, J.M., Riemer, M.F., Sancio, R.B., Bray, J.D., Kayen, R.E. and Faris, A. (2003) Recent advances in soil liquefaction engineering: a unified and consistent framework. *Proceedings of the 26th Annual ASCE Los Angeles Geotechnical Spring Seminar, Keynote Presentation*, H.M.S. Queen Mary, Long Beach, California. 71 pp.

Van der Poel, J.T. and de Groot, M.B. (1998) Cyclic load tests on a caisson breakwater placed on sand. *Proceedings of the International Conference Centrifuge 98* 1, 403 – 408

Wood, D.M. (1990) *Soil Behaviour and Critical State Soil Mechanics.* Cambridge University Press, Cambridge.

Youd, T.L. and Idriss, I.M. (2001) Liquefaction resistance of soils: summary report from the 1996 NCEER and 1998 NCEER/NSF workshops on evaluation of liquefaction resistance of soils. *ASCE Journal of Geotechnical and Geo-Environmental Engineering* 127 (10), 297 – 313.

Zhang, X.Y., Lee, F.H. and Leung, C.F. (2009) Response of caisson breakwater subjected to repeated impulsive loading. *Géotechnique* 59 (1), 3 – 16.

PART III

EXCAVATIONS

Chapter 6

Braced Excavation Collapse: Nicoll Highway, Singapore

TABLE OF CONTENTS

A.M. Puzrin et al., *Geomechanics of Failures*, DOI 10.1007/978-90-481-3531-8_6,

Chapter 6
Braced Excavation Collapse: Nicoll Highway, Singapore

6.1 Case Description

The island city-state of Singapore has one of the best public transport systems in the world. The Mass Rapid Transit (MRT) railway network is the main backbone of this system. Construction of a new 33.6 km long underground Circle MRT Line (CCL) started in 2001. Costing approximately $6.7 billion, the CCL will link all the radial lines leading to the city (Fig. 6.1a). The first stage of this construction (CCL1) was 5.4 km long and was subdivided into two contracts: C824 and C825 (Fig. 6.1b).

(a)

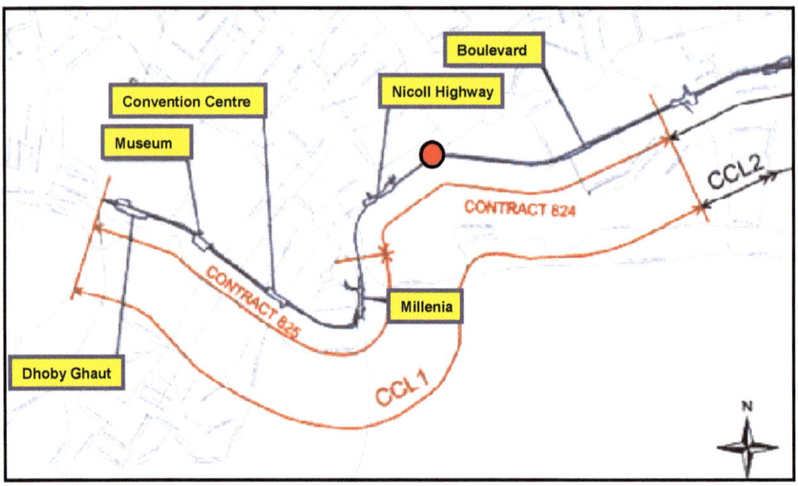

(b)

Figure 6.1 The Circle Line (CCL): (a) location on the island (Wikipedia, 2009); (b) the first stage (after the COI Report, 2005).

The scope of the design-and-build contract C824 comprised of Nicoll Highway and Boulevard Stations, 800 m of bored tunnels, and 1,600 m of cut and cover tunnels. The main contractor for these works was the Nishimatsu-Lum Chang Joint Venture. The client was the Land Transport Authority of Singapore.

Around 3:30 pm on April 20, 2004, the 33.5 m deep cut and cover excavation required for the construction of tunnels between Nicoll Highway and Boulevard stations (the red circle in Figures 6.1a;b) collapsed. The collapse was catastrophic, resulting in four fatalities and a lot of damage (Fig. 6.2a;b), which delayed completion of the CCL1 stage by about four years.

(a)

(b)

Figure 6.2 The Nicoll Highway cut and cover excavation (after the COI Report, 2005): (a) before; and (b) after the collapse.

The Singapore Government appointed a Committee of Inquiry (COI) to ascertain the causes and circumstances of the incident. The COI Report (2005), submitted to the Government in May 2005 and made public on the Internet, is the main source of information for this chapter.

The COI came to a conclusion, which in our computer-dominated age sounds like a paradox: a wrong use of numerical modeling in geotechnical design, together with some structural errors, were the main causes of the Nicoll Highway collapse.

6.1.1 Design and construction

The cut and cover tunnels for the contract C824 were constructed using the "bottom-up" method (Fig. 6.3). The diaphragm walls used at this section of the cut and cover tunnels were 0.8 m thick and penetrated 1 – 3 m into stiffer strata at the depth of about 40 – 45 m. Two layers of jet grout slabs, constructed using interlocking jet grout piles (JGP), were also installed after construction of the slurry walls to minimize deflection and ground movement during the excavation work. The upper 1.5 m thick layer of JGP (temporary – to be removed before the tenth strut level installation) was placed at about 28 m below ground level. The lower (permanent) 2.6 m thick JGP layer was built 33.5 m below the ground level. The excavation was kept open by a strutted system comprising of steel king posts, walers, and 10 levels of struts. Struts were spaced at 3.0 – 3.5 m vertically and 4.0 m horizontally. The width of excavation was about 20 m and it was about 33.5 m deep.

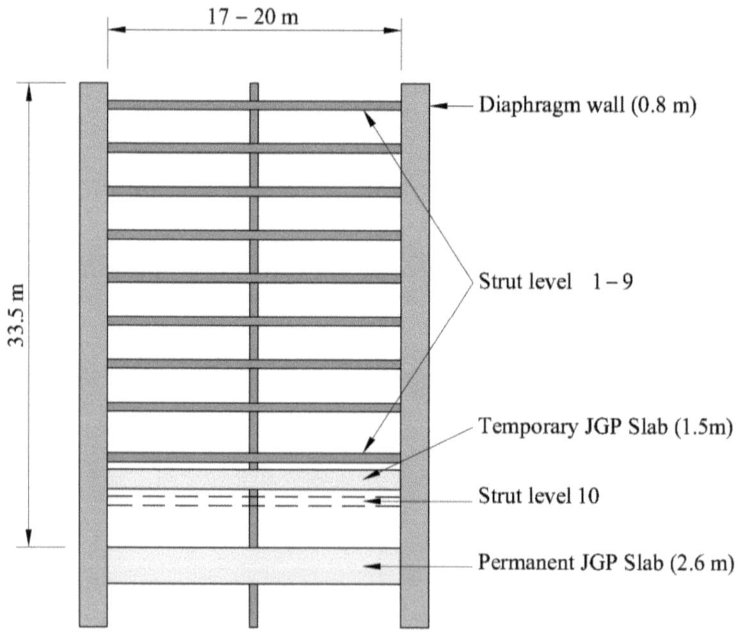

Figure 6.3 The excavation retaining structures (after the COI Report, 2005).

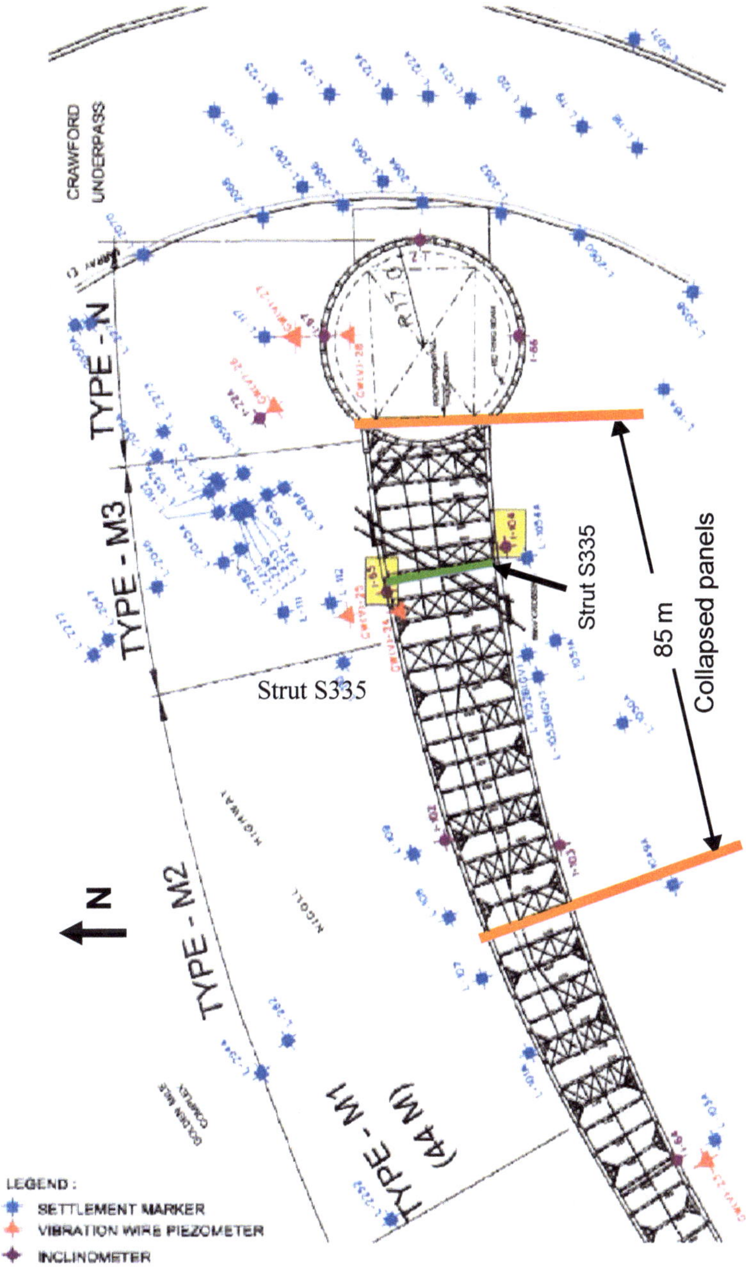

Figure 6.4 The monitoring system (after the COI Report, 2005): inclinometers I-65 and I-104 marked yellow; strut S335 marked green.

An extensive monitoring system was installed (Fig.6.4). It included more than 2000 monitoring instruments: settlement markers and inclinometers to measure soil and wall displacements; vibrating wire piezometers to measure pore water pressures; strain gages and load cells to measure the strut loads. The readings from the inclinometers I-65 and I-104 (marked yellow in Fig. 6.4) and strain gages and load cells in the strut S335 (marked green in Fig. 6.4) are crucial for understanding the cause of excavation collapse.

6.1.2 The collapse

The COI Report states: "Warnings of the approaching collapse were present from an early stage but these were not taken seriously". Indeed, as early as in August, 2003, it became clear that something was wrong with the design. First, in a launch shaft for the tunnel boring machines, across the river east from the collapsed area, measured wall deflection exceeded 500 mm, while the calculated design level was only 190 mm. This caused damage to the retaining walls, ground settlements and cracks at a cricket field nearby.

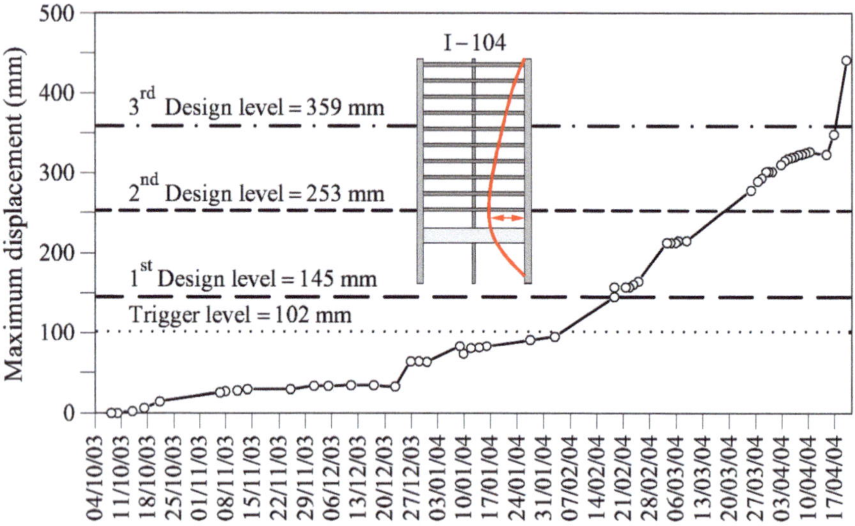

Figure 6.5 Displacement of the southern wall in time (after the COI Report, 2005): measurements for inclinometer I-104.

Next, in January 2004, design deflection level was exceeded and cracks appeared in the retaining structures in the cut and cover excavation west of the collapsed area. The design deflection level had to be revised upwards from 222 to 522 mm (from 125 to 313 mm closer to the collapse area). This was accompanied by numerous complaints received from owners of neighboring properties, including ground settlements of a driveway, cracks in buildings and other building damages. Manpower resources from the C824 project team were deployed to handle the repair works.

Finally, in the collapsed area itself, the inclinometer I-104 indicated that the original design level of 145 mm in the southern diaphragm wall was exceeded as early as in February, 2004 (Fig. 6.5). At that moment installation of the sixth strut level was in progress.

(a)

(b)

Figure 6.6 Buckling of walers (after the COI Report, 2005): (a) a stiffener plates; (b) C-channels.

This deflection was accompanied by buckling of waler stiffener plates (Fig. 6.6a), which were replaced by C-channels (Fig. 6.6b) starting from the seventh level. On a basis of back analysis it was decided to update the design level to 253 mm and to increase the preloading of the struts from 50 to 70%.

Unfortunately, this did not help, and in the end of March 2004, the new design level was exceeded again (Fig. 6.5). Another back analysis produced another revision of the design level up to 359 mm which, nevertheless, when the last measurement was taken on the day of collapse, was exceeded by as much as 22% (441 mm)!

On the day of collapse, April 20, 2004, the ninth level of struts was completed and the excavation proceeded to the final tenth level. In order to achieve that, the first (temporary) layer of JGP had to be excavated. In the morning, the workers were carrying out excavation work and preparing to install the tenth level of struts. At around 8.30 am, they heard "thung" sounds at the excavation base (frequency of about 4 – 5 times within 10 minutes). These were waler beams buckling.

In order to strengthen the waler beams, it was decided to pour concrete into their top section and to weld stiffener plates into their bottom section. This did not help – the beams were already too distorted to hold the concrete. The frequency of the "thung" sounds increased. Another contingency measure – pouring lean concrete into the excavation base to replace the excavated JGP slab – came too late. At about 3.30 pm, a section of the temporary diaphragm wall supporting the excavation adjoining the Nicoll Highway Station box had collapsed.

This caused soil subsidence of the immediate surrounding area, which propagated rapidly, covering approximately 100×130 m area and sucking in heavy equipment, site offices, etc. (Fig. 6.7). This also resulted in substantial damage to about 100 m section of the Nicoll Highway carriageway adjacent to the abutment of Merdeka Bridge. The incident had resulted in four fatalities (three construction workers and one engineer) and three injuries. Fortunately, no motorists were traveling along that section of the Nicoll Highway when the incident happened, otherwise the number of deaths and injuries could have been even higher.

6.1.3 The problem

The site investigation indicated that the soil profile can be divided into the following strata (Fig. 6.8a): Fill (the area was reclaimed 20 – 50 years ago, the water level is 2 m below the surface); Estuarine (organic) deposits; Upper and Lower Marine Clays; and more of Fluvial and Estuarine deposits. These soft clays are underlain by stiffer sands, silts, and clays of the Old Alluvium. The best estimate of the undrained shear strength from piezocone tests (CPT) is given in Figure 6.8b.

The Committee of Inquiry established two main causes of the collapse. One is geotechnical – a wrong estimate of the undrained earth pressure acting on the retaining structures. Another is structural – an error in design of the bearing capacity of the walers, unable to take redistributed loads once one of them failed.

(a)

(b)

Figure 6.7 The progress of the collapse (after the COI Report, 2005): (a) 3:33 pm – the second crane is standing; (b) 3:34 pm – the second crane collapsed, resulting in the death of its operator.

Combined together these errors resulted in a considerable under-design leading to the failure of the retaining structures. Replacement of the stiffener plates (Fig. 6.6a) by C-channels (Fig. 6.6b) changed the mode of failure from ductile to brittle (catastrophic).

Other significant contributing causes were: wrong use of the geotechnical back analysis and inefficiency of instrumentation and monitoring systems. Additional contributing causes could have been insufficient toe-penetration of the diaphragm wall into the impermeable strata; lower than expected JGP strength; long times at which excavation was left unsupported; no strut failure check in the back analysis; etc. In this chapter, however, we focus on the main geotechnical aspect of the Nicoll Highway Collapse – a wrong estimate of the undrained shear strength and, as a result, of the lateral earth pressure.

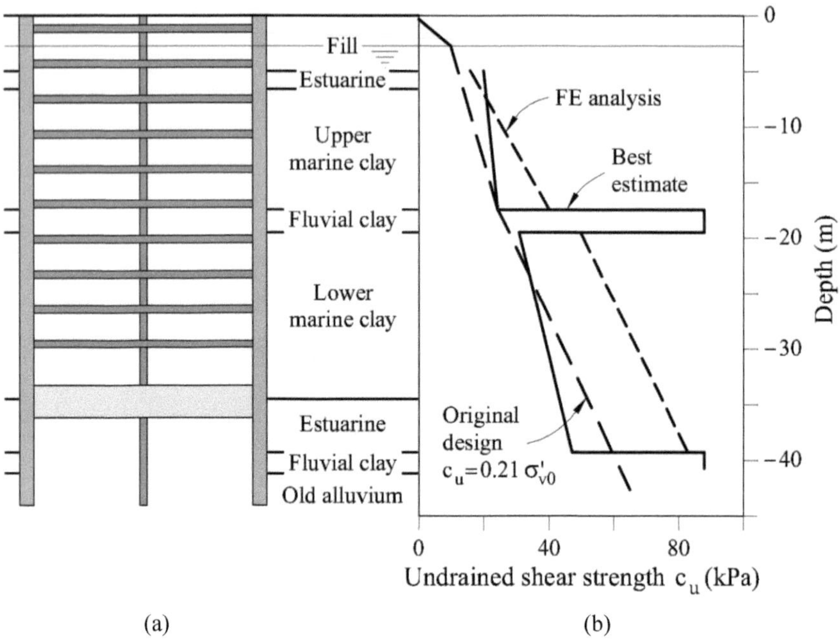

Figure 6.8 Soil profile (Whittle and Davies, 2006): (a) classification; (b) undrained shear strength (the best estimate, the original design and the FE analysis values).

6.1.4 The undrained earth pressure analysis

A rapid temporary excavation in saturated clay does not leave time for excess pore water pressures to dissipate. This means that the geotechnical earth pressure analysis has to be performed using the assumption of undrained loading. This kind of analysis is easily performed in total stresses using undrained shear strength c_u. It can also be performed in effective stresses, with the effective strength parameter φ'. But in the latter case, the excess pore water pressures, generated due to shearing in this kind of normally to slightly overconsolidated Marine Clays, have to be properly modeled.

The original design used simple active and passive pressure formulas based on the total stress analysis with the values of c_u close to those from the best estimate of the field test data (Fig. 6.8b). A requirement, however, was to use one of the commercially available Finite Element codes. The designers have chosen PLAXIS, although lacking sufficient experience in its application. The Mohr – Coulomb model is built into PLAXIS and the program manual recommended the use of the effective strength parameter φ' for undrained analysis, which was followed by the designer. As a result, the undrained shear strength was effectively overestimated by almost 50% (Fig. 6.8b)! How could that happen and why was it not noticed?

The Mohr – Coulomb model assumes isotropic elastic behavior during pre-failure deformation. Therefore, in a drained case, pure shear would not cause any volumetric strains. In undrained shear loading, when no volume change is allowed, this results in no pore water pressure being generated. The mean effective stress p' stays constant and the corresponding effective stress path to failure in $p' - q$ triaxial stress space is vertical (Fig. 6.9a).

In reality, however, normal to slightly overconsolidated clays have a tendency to contract during drained loading. Since the water is incompressible in undrained shearing, this results in positive excess pore water pressures. The corresponding effective stress path to failure in $p' - q$ triaxial stress space curves to the left from vertical (Fig. 6.9b), and hits the failure envelope at a much lower deviatoric failure stress q_f, related to the undrained shear strength via $q_f = 2c_u$.

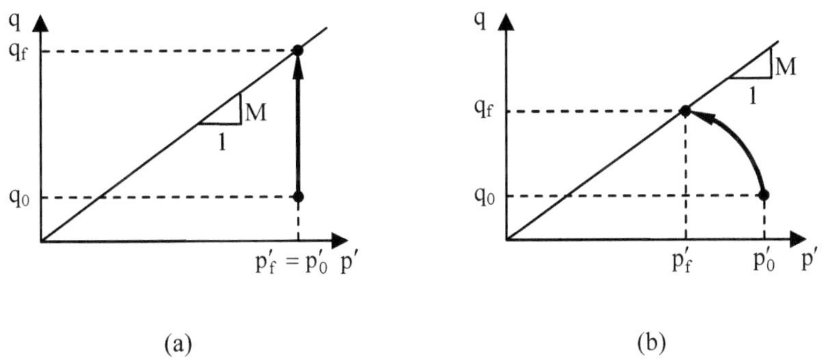

Figure 6.9 Undrained effective stress path: (a) Mohr-Coulomb model; (b) normally consolidated clay.

6.2 The Theory

6.2.1 Long-term earth pressures

The long term stability of a supported excavation is assessed using an assumption of drained shearing, when all the excess pore water pressures have dissipated. In a lack of cohesion (typical for normally consolidated clays), effective active earth pressure (Fig. 6.10a) acting at the depth H on a smooth vertical retaining wall, can be calculated as

$$e_a = K_a \gamma' H, \tag{6.1}$$

where

$$K_a = \frac{1-\sin\varphi'}{1+\sin\varphi'} \tag{6.2}$$

is the active earth pressure coefficient; γ' is the effective specific weight of soil; and φ' is the effective angle of internal friction.

In a braced excavation, displacements are not large enough for an active failure state to develop. In this case, according to Terzaghi *et al.* (1996), the effective force will be 30% larger than in the free standing wall. Also, the earth pressure diagram becomes rectangular (Fig. 6.10b).

Naturally, the stability of the wall depends on the total and not effective earth pressures. To calculate the former, the pore water pressures u have to be added to the effective earth pressures calculated using Equation (6.1). In the long term, however, these pore water pressures are likely to be hydrostatic.

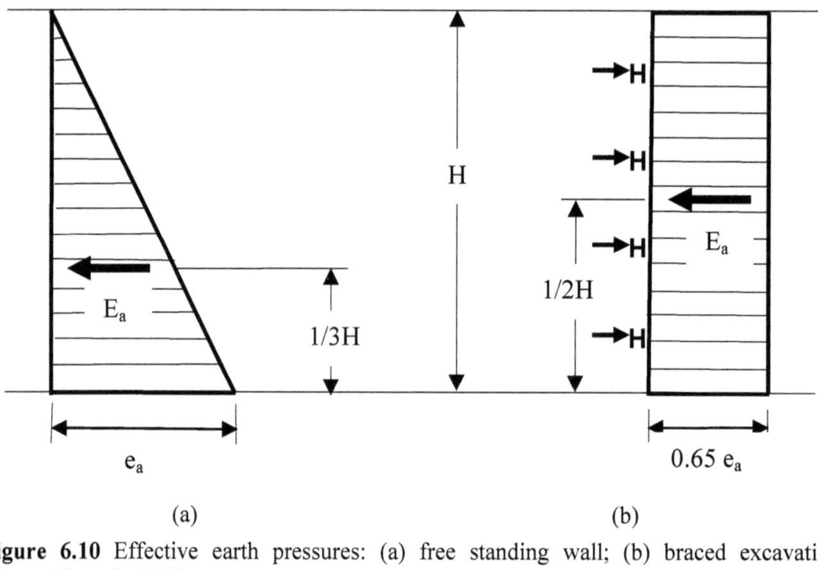

Figure 6.10 Effective earth pressures: (a) free standing wall; (b) braced excavation (Terzaghi *et al.*, 1996).

6.2.2 Short-term earth pressures

In saturated clays, the short-term stability of a supported excavation is assessed using an assumption of undrained shearing. In normally to slightly overconsolidated clays, this undrained shearing generates positive, while in strongly overconsolidated clays, negative excess pore water pressures, which do not have sufficient time to dissipate during the excavation. In this case, the analysis is normally carried out in total stresses. The diagram of the total active earth pressure acting on a wall of a braced excavation of depth H in normally consolidated clay, is shown in Figure 6.11a (Terzaghi *et al.*, 1996). The total active earth pressure is given by

$$e_a = K\gamma H, \tag{6.3}$$

where

$$K = 1 - m\frac{4c_u}{\gamma H}, \tag{6.4}$$

$m \approx 0.8$ for normally to slightly overconsolidated clays; γ is the total specific weight of soil; c_u is the undrained shear strength. The total active earth pressure acting on a wall of a braced excavation of depth H in overconsolidated clay is shown in Figure 6.11b.

In principle, the short-term stability analysis can also be performed in effective stresses. In this case, however, it is necessary to be able to correctly calculate the excess pore water pressures generated in soil due to undrained shearing. We need these pore water pressures in order to correctly compute the total stresses, but, even more importantly – to properly assess the undrained shear strength of soil.

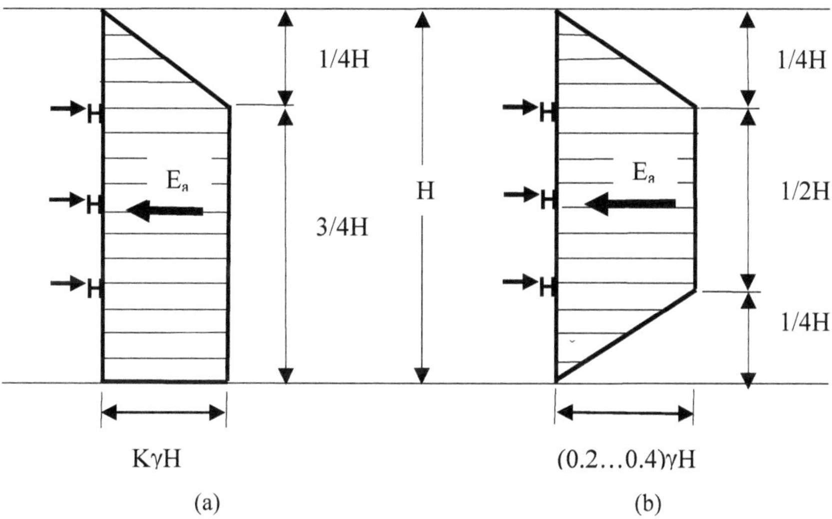

Figure 6.11 Total earth pressures (Terzaghi *et al.*, 1996) for short-term stability of a braced excavation in: (a) normally consolidated; and (b) overconsolidated clay.

6.2.3 The undrained shear strength

The undrained shear strength of clay c_u can be derived in the laboratory from unconsolidated undrained (UU) triaxial tests, or in situ from vane tests or CPT (cone penetration tests). In normally consolidated clays, due to development of the positive excess pore water pressures during undrained shearing, the effective stresses decrease (Fig. 6.9b). Therefore, this experimentally determined c_u appears to be significantly smaller than the clay strength at the same initial effective stresses in drained shearing (slow loading). An empirical rule of thumb gives the following approximate relationship for the undrained shear strength of normally consolidated clays:

$$c_u = k\sigma'_{v0}, \qquad k = 0.21 - 0.25, \tag{6.5}$$

where σ'_{v0} is the in situ geostatic vertical effective stress. The lower values of $k = 0.21$ are valid for soft marine clays, which is the case in Singapore. The shear strength values for the original design in Figure 6.8b were, apparently based on $c_u = 0.21\sigma'_{v0}$. Using the effective stress approach, different constitutive models predict the relationship (6.5) with a different degree of accuracy.

6.2.4 The Mohr – Coulomb model

The Mohr – Coulomb model assumes isotropic elastic behavior during pre-failure deformation. In undrained triaxial shear loading, when no volume change is allowed, this results in no pore water pressure being generated. The mean effective stress p' stays constant and equal to the initial mean effective stress p'_0, and the corresponding effective stress path to failure in $p' - q$ triaxial stress space is vertical (Fig. 6.9a). The deviatoric stress at failure q_f can be then calculated as

$$q_f = M\,p'_f = M\,p'_0, \tag{6.6}$$

where

$$M = \frac{6\sin\varphi'}{3 - \sin\varphi'}, \tag{6.7}$$

$$p'_0 = \frac{1 + 2K_0}{3}\sigma'_{v0}, \tag{6.8}$$

where K_0 is the at-rest earth pressure coefficient, which for normally consolidated clays, can be estimated using the formula (Jaky, 1944)

$$K_0 = 1 - \sin\varphi'. \tag{6.9}$$

Substituting expressions (6.7) – (6.9) into (6.6) and noting that

$$c_u = \frac{\sigma_{1f} - \sigma_{3f}}{2} = \frac{q_f}{2}, \tag{6.10}$$

we obtain the undrained shear strength estimate

$$c_u = k\sigma'_{v0}, \tag{6.11}$$

where

$$k = k_{MC} = \sin\varphi' \frac{3-2\sin\varphi'}{3-\sin\varphi'}. \tag{6.12}$$

For a typical range of $\varphi' = 20^\circ - 24^\circ$: $k \approx 0.30 - 0.35$, which is $40-50\%$ higher than the value in relationship (6.5)!

6.2.5 The Modified Cam Clay model

The Modified Cam Clay model (Burland and Roscoe, 1968), in contrast, assumes plastic behavior in pre-failure deformation using an elliptical yield surface (Fig. 6.12a):

$$\frac{q^2}{M^2 p'^2} + 1 = \frac{p'_c}{p'}, \tag{6.13}$$

where p'_c is the pre-consolidation pressure, serving as a hardening parameter related to the specific volume via the equation of the virgin compression line (VCL, Fig. 6.12b):

$$V_c = N - \lambda \ln p'_c. \tag{6.14}$$

The failure takes place when the stress state (p'_f, q_f) reaches the critical state line (CSL), which in the triaxial stress space (Fig. 6.12a) is identical to the Mohr – Coulomb failure envelope

$$q_f = M p'_f. \tag{6.15}$$

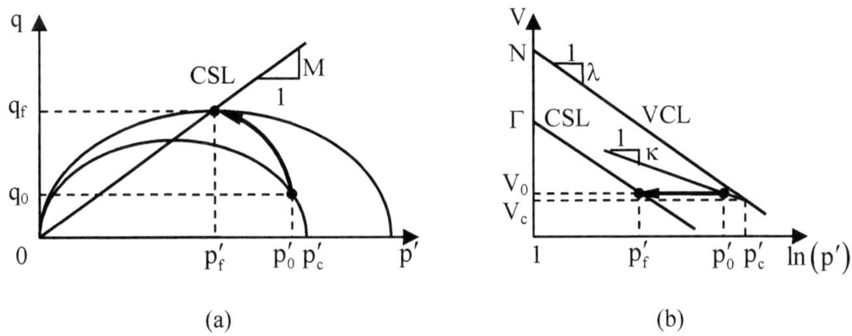

Figure 6.12 Modified Cam Clay model: (a) undrained effective stress path; (b) constant volume (undrained) deformation.

In the volumetric stress-deformation space (Fig. 6.12b), the CSL is parallel to the VCL:

$$V = \Gamma - \lambda \ln p'_f , \tag{6.16}$$

where

$$\Gamma = N - (\lambda - \kappa) \ln 2 \tag{6.17}$$

and κ defines elastic volumetric behavior (Fig. 6.12b):

$$V = V_c - \kappa \ln(p'/p'_c) . \tag{6.18}$$

The initial stress state after one-dimensional consolidation is given by

$$p'_0 = \frac{1+2K_0}{3} \sigma'_{v0} , \qquad q_0 = (1 - K_0) \sigma'_{v0} . \tag{6.19}$$

The corresponding initial volume is found from Equations (6.14) and (6.18):

$$V_0 = N - \lambda \ln p'_c - \kappa \ln p'_0 / p'_c , \tag{6.20}$$

where p'_c / p'_0 for the initial stress state is found from Equation (6.13):

$$\frac{p'_c}{p'_0} = \frac{q_0^2}{M^2 p_0'^2} + 1 . \tag{6.21}$$

During the undrained loading to failure, the initial volume V_0 does not change (Fig. 6.12b), therefore we obtain from Equation (6.16)

$$p'_f = \exp[(\Gamma - V_0)/\lambda] \tag{6.22}$$

which, after substitution of (6.17) and (6.20), turns into

$$p'_f = p'_0 \left(\frac{1}{2} \frac{p'_c}{p'_0} \right)^{1-\kappa/\lambda} . \tag{6.23}$$

Substitution of Equations (6.21) and (6.23) into (6.15) gives the deviatoric stress at failure:

$$q_f = M \, p'_0 \left(\frac{1}{2} \frac{q_0^2}{M^2 p_0'^2} + \frac{1}{2} \right)^{1-\kappa/\lambda} , \tag{6.24}$$

which, using Equations (6.7), (6.9), (6.10) and (6.19), can be transformed into the following relationship for the undrained shear strength:

$$c_u = k \sigma'_{v0} , \tag{6.25}$$

$$k = k_{MCC} = k_{MC}\left(\frac{1}{8}\frac{\sin^2\varphi'}{k_{MC}^2}+\frac{1}{2}\right)^{1-\kappa/\lambda}, \qquad (6.26)$$

where k_{MC} is calculated from Equation (6.12). For a typical range of values of $\varphi' = 20^\circ - 24^\circ$ and $\kappa/\lambda = 0.2$, we obtain $k = 0.215 - 0.251$, i.e. very close to that in relationship (6.5)!

The effective undrained stress path is curved to the left (Fig. 6.12a) and given by

$$\frac{q^2}{M^2 p'^2}+1=\left(\frac{p'_0}{p'}\right)^{\frac{1}{1-\kappa/\lambda}}\left(\frac{q_0^2}{M^2 p_0'^2}+1\right). \qquad (6.27)$$

6.2.6 The Original Cam Clay model

The undrained shear strength can be derived for the original Cam Clay (Schofield and Wroth, 1968; Muir Wood, 1996) following the above procedure and simply replacing Equation (6.12) of the yield surface and expression (6.17) for parameter Γ by

$$\frac{q}{M\,p'} = \ln\frac{p'_c}{p'} \qquad (6.28)$$

and

$$\Gamma = N - (\lambda - \kappa), \qquad (6.29)$$

respectively.

The corresponding deviatoric stress at failure,

$$q_f = M\,p'_0 \exp\left[\left(\frac{q_0}{M\,p'_0}-1\right)\left(1-\frac{\kappa}{\lambda}\right)\right], \qquad (6.30)$$

can be transformed using Equations (6.7), (6.9), (6.10) and (6.19) into the following relationship for the undrained shear strength:

$$c_u = k\sigma'_{v0}, \qquad k = k_{OCC} = k_{MC}\exp\left[\left(\frac{\sin\varphi'}{2k_{MC}}-1\right)\left(1-\frac{\kappa}{\lambda}\right)\right], \qquad (6.31)$$

where k_{MC} is calculated from Equation (6.12). For a typical range of values of $\varphi' = 20^\circ - 24^\circ$ and $\kappa/\lambda = 0.2$, we obtain $k = 0.212 - 0.248$, i.e. again close to that in relationship (6.5)!

The effective undrained stress path is curved to the left and given by

$$\frac{q}{M\,p'} = \frac{\ln\left(p_0'/p'\right)}{1-\kappa/\lambda} + \frac{q_0}{M\,p_0'} \,. \tag{6.32}$$

6.2.7 Summary

For stability analysis of braced excavations in saturated normally consolidated clays, it is necessary to determine earth pressures under undrained loading conditions. Total stress analysis is probably the easiest way to calculate these pressures. The effective stress analysis will, in principle, produce the same result, provided the chosen soil model is capable of accurate prediction of the undrained shear strength. We have shown that the Mohr – Coulomb model overpredicts the undrained shear strength by as much as 50%. In contrast, both the original and modified Cam Clay models produce very close predictions, both within 1% of the experimental data, in spite of the fact that their analytical expressions look very different.

6.3 The Analysis

The theory presented above provides tools for a simplified calculation of the short-term stability of the Nicoll Highway excavation, with the purpose of demonstrating how the overestimation of c_u could lead to its collapse.

6.3.1 The simplified model

Soil parameters defined from the laboratory and field tests and used in the original design are given in Table 6.1. For a simplified stability analysis, we assume that the soil profile is built of two layers only – the Upper and the Lower Marine Clays, with thicknesses of 19.2 and 12.8 m, respectively (Fig. 6.13). The clay in both layers is assumed to be normally consolidated.

Table 6.1 Soil parameters used in the original design (after the COI Report, 2005).

Stratum	Symbol	SPT-N	Bulk density, γ (kN/m³)	c_u (kN/m²)	c' (kN/m²)	ϕ' (°)	Elastic modulus (kN/m²)	Permeability, k (m/s)
Fill	F	6 ± 3	19	25	0	30	10,000	10^{-6}
Estuarine	E	2 ± 2	15	15 (0 to 10 m)	0	18	6,000	10^{-9}
Upper Marine	M (upper)	$0-1$	16	10 (0 to 5 m) 20 (5 to 15 m)	0	22	4,000 8,000	10^{-9}
Fluvial Clay	F2	10 ± 5	19	20 (0 to 10 m)	0	22	8,000	10^{-9}
Lower Marine	M (Lower)	$0-1$	16	20+1.6 (z-15)	0	24	400 c_u	10^{-9}
Estuarine	E	2 ± 2	15	15+2.3 (z-10)	0	18	400 c_u	10^{-9}
Fluvial Clay	F2	10 ± 5	19	20+ (z-10)	0	22	400 c_u	10^{-9}
Old Alluvium	OA (W)	19 ± 6	20	5N	0	32	2,000 N	5×10^{-7}
	OA (SW-2)	40 ± 6	20		5	32		5×10^{-7}
	OA (SW-1)	71 ± 12	20		5	33		5×10^{-8}
	OA (CZ)	>100	20		10	35		5×10^{-8}
JGP-1	JGP-1		16	300			15,000	10^{-10}

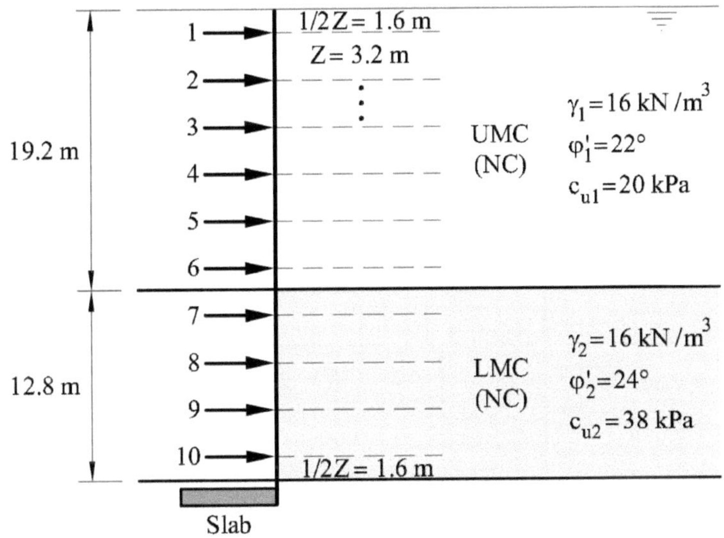

Figure 6.13 Simplified geotechnical model of the collapsed excavation.

The groundwater level is conservatively assumed to be at the soil surface. For simplicity, the soil parameters for each layer in Figure 6.13 have been assumed to be constant with depth by averaging the values from Table 6.1. The undrained shear strength of $c_u = 20\,\text{kPa}$ and $c_u = 38\,\text{kPa}$, for the Upper and Lower Marine Clay, respectively, are the conservative values taken from the best estimate of the field and laboratory data in Figure 6.8b and Table 6.1. Note that, for the Lower Marine clay, the best estimate of the field tests in Figure 6.8b is lower that the original design estimate of $c_u = 0.21\sigma'_{v0}$ for normally consolidated soft marine Clays, which was based on the assumption of the hydrostatic pore water pressure. This is because the Lower Marine Clay is still undergoing consolidation from the land reclamation of 1970s and is, therefore, underconsolidated.

The excavation is assumed to be supported by 10 strut levels, spaced vertically at equal intervals of 3.2 m. The first strut level was located 1.6 m below the surface. The surface surcharge is assumed to be equal to zero. The last, tenth strut level was built after the temporary JGP slab at the same level was removed.

6.3.2 The long-term stability

The effective earth pressures for the long term stability analysis are determined using Figure 6.14, which is a two-layer equivalent of the diagram in Figure 6.10b. The total earth pressure is calculated by adding the water pressures which are assumed to be hydrostatic. The corresponding strut loads at the end of the excavation are calculated in Table 6.2 by integrating earth pressures over the length of influence of each strut, with the total active load being $E_a = 6{,}884\ \text{kN/m}$.

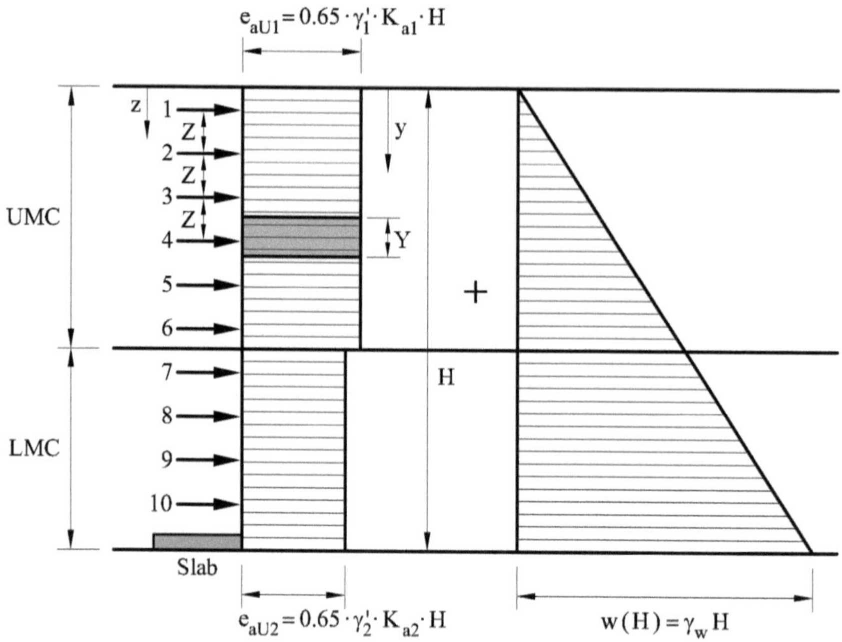

Figure 6.14 The active earth pressure for the long-term stability analysis.

Table 6.2 Strut loads for the long-term stability analysis.

Strut loads (Project design)

Strut level	Schicht	z (m)	Z (m)	y (m)	Y (m)	e_{aU} (kN/m²)	w(z) (kN/m²)	Strut load with JGP Slab FS = 1.0 (kN/m)	Design (kN/m)
0	UM	0.00							
	UM		1.60						
1	UM	1.60			3.20	56.78	16.00	232.89	568.00
	UM		3.20	3.20					
2	UM	4.80			3.20	56.78	48.00	335.29	1,018.00
	UM		3.20	6.40					
3	UM	8.00			3.20	56.78	80.00	437.69	1,816.00
	UM		3.20	9.60					
4	UM	11.20			3.20	56.78	112.00	540.09	1,635.00
	UM		3.20	12.80					
5	UM	14.40			3.20	56.78	144.00	642.49	1,458.00
	UM		3.20	16.00					
6	UM	17.60			3.20	56.78	176.00	744.89	1,322.00
	UM		3.20	19.20					
7	LM	20.80			3.20	52.63	208.00	834.02	2,130.00
	LM		3.20	22.40					
8	LM	24.00			3.20	52.63	240.00	936.42	2,332.00
	LM		3.20	25.60					
9	LM	27.20			3.20	52.63	272.00	1,038.82	2,173.00
	LM		3.20	28.80					
10	LM	30.40			3.20	52.63	304.00	1,141.22	
	LM		3.20	32.00					
$Slab_B$	LM	33.60							
								6,883.85	14,752.00

Here and below, z indicates the depth of each strut and y indicates the depth of the two boundaries of the length of influence for each strut.

In Figure 6.15 the calculated strut loads are compared with unfactored strut loads used in the design with a total strut load of $E_{\text{struts}} = 14,752$ kN/m. The safety factor for the long-term stability is, therefore:

$$F_s = \frac{E_{\text{struts}}}{E_a} = \frac{14,752}{6,884} = 2.14. \quad (6.33)$$

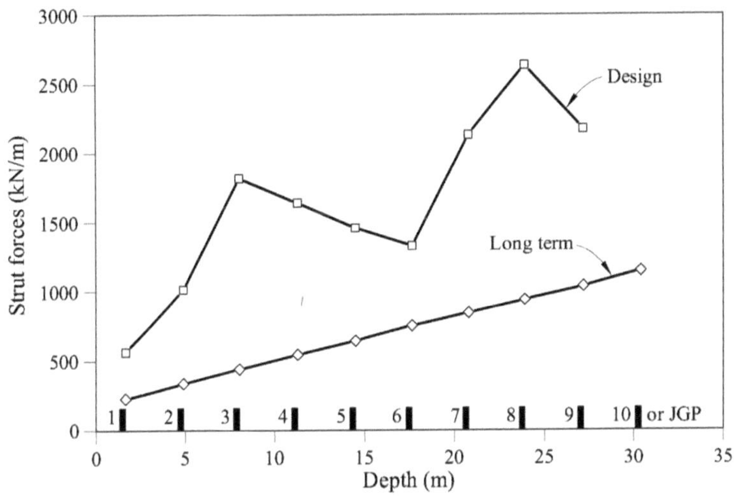

Figure 6.15 Strut loads for the long-term stability analysis compared to the design.

6.3.3 The short-term stability

The total earth pressures for the short-term stability analysis are determined using Figure 6.16, which is a two-layer equivalent of the diagram in Figure 6.11a. The corresponding strut loads at the end of the excavation are calculated in Table 6.3, with the total active load being E_a = 11,717 kN/m.

In Figure 6.17, these strut loads are compared with the long-term strut loads and with the unfactored strut loads used in the design. Clearly, the short-term stability is of greater concern here and should serve as a basis for the design. The design loads at the lower strut levels are significantly higher, because they take into account the redistribution of forces during the removal of the temporary JGP slab and struts for the tunnel construction. The safety factor for the short-term stability is

$$F_s = \frac{E_{\text{struts}}}{E_a} = \frac{14,752}{11,717} = 1.26. \quad (6.34)$$

Though significantly lower than for long-term stability case, this safety factor alone still does not explain the collapse. This explanation requires an understanding of the excavation process, redistribution of strut forces, and of progressive failure.

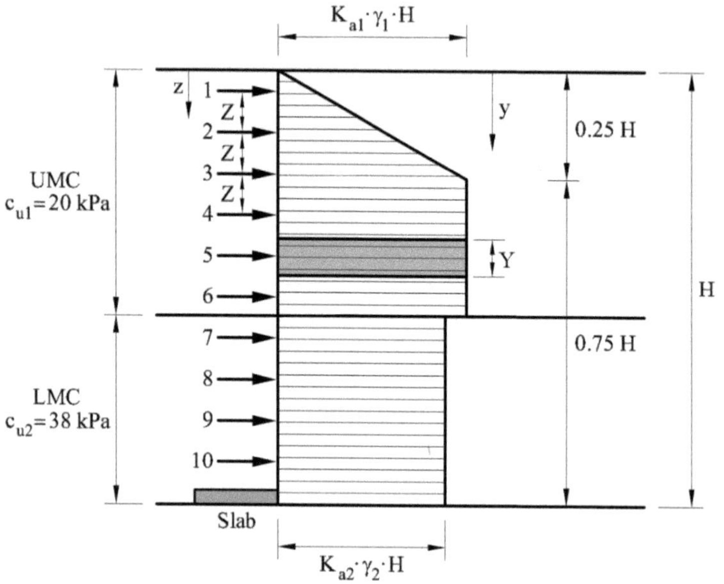

Figure 6.16 Active earth pressure diagram for the short-term stability analysis.

Table 6.3 Strut loads for the short-term stability analysis.

Strut loads (real) **0.25H=8.40**

Strut level	Layer	z (m)	Z (m)	y (m)	Y (m)	γ (kN/m³)	m (-)	c_u (kN/m²)	K_a (-)	$e_a=K\cdot\gamma\cdot D$ (kN/m²)	Strut load calc. FS=1 (kN/m)
0	UM	0.00									
	UM		1.60			16.00	0.80	20.00			
1	UM	1.60			3.20						286.72
	UM		3.20	3.20		16.00	0.80	20.00	0.88	179.20	
2	UM	4.80			3.20						860.16
	UM		3.20	6.40		16.00	0.80	20.00	0.88	358.40	
3	UM	8.00			3.20						1,272.32
	UM		3.20	9.60		16.00	0.80	20.00	0.88	448.00	
4	UM	11.20			3.20						1,433.60
	UM		3.20	12.80		16.00	0.80	20.00	0.88	448.00	
5	UM	14.40			3.20						1,433.60
	UM		3.20	16.00		16.00	0.80	20.00	0.88	448.00	
6	UM	17.60			3.20						1,433.60
	UM		3.20	19.20		16.00	0.80	20.00	0.88	448.00	
7	LM	20.80			3.20						1,249.28
	LM		3.20	22.40		16.00	0.80	38.00	0.76	390.40	
8	LM	24.00			3.20						1,249.28
	LM		3.20	25.60		16.00	0.80	38.00	0.76	390.40	
9	LM	27.20			3.20						1,249.28
	LM		3.20	28.80		16.00	0.80	38.00	0.76	390.40	
10	LM	30.40			3.20						1,249.28
	LM		3.20	32.00		16.00	0.80	38.00	0.76	390.40	
Slabs	LM	33.60									
											11,717.12

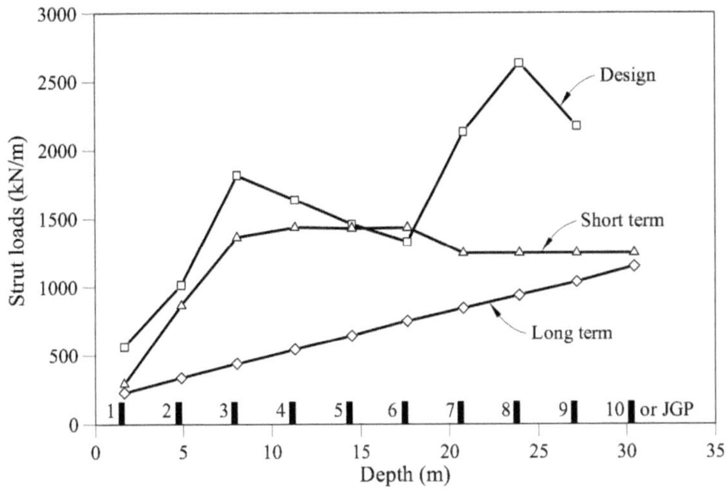

Figure 6.17 Strut loads for the short-term stability analysis compared to the design.

6.3.4 Excavation progress and collapse

Evolution of the total earth pressures with the progress of the excavation is shown in Figure 6.18, which applies the diagram from Figure 6.16 for stepwise increasing depth. The corresponding strut loads at various stages of excavation are shown in Figure 6.19, where they are compared with the strut loads used in the design.

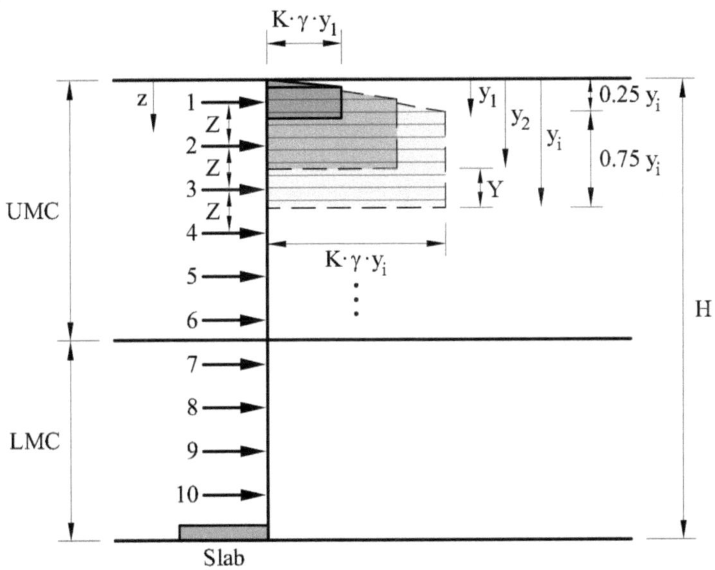

Figure 6.18 Evolution of the active earth pressure diagram for the short-term stability analysis with the progress of the excavation.

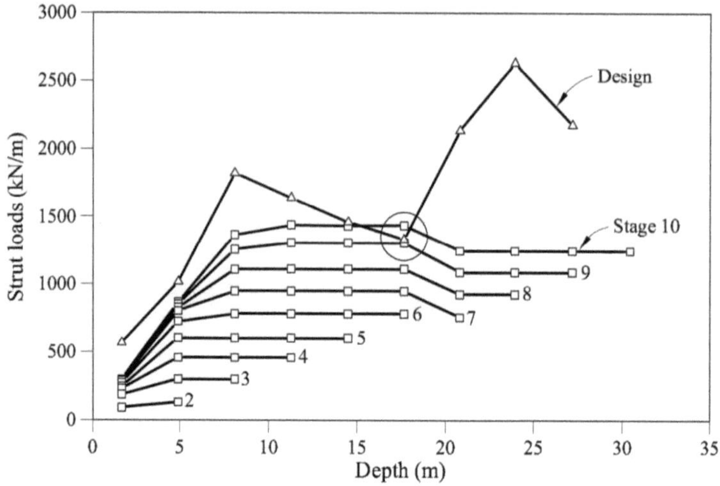

Figure 6.19 The strut loads for the short-term stability analysis at different excavation stages.

It can be seen that, as the excavation progressed, the load in the sixth strut level approached the design load, and this was exactly the location where the structural buckling problems (Fig. 6.6) had been encountered already in February 2004 – two months before the collapse.

The critical stage of excavation – removal of the temporary JGP slab before the installation of the tenth strut level – is shown in Figure 6.20a. Here the strut loads before and after the removal of the temporary JGP slab are compared to the design strut loads. It can be seen that, as the slab was removed and if its full load were transferred to the ninth strut level, the design load in the ninth level would be exceeded by 15%. Together with the structural error in the waler beam design, this would lead to the ninth level failure. The load then would be redistributed to the eighth level struts (as was indeed measured, see Fig. 6.20b), which would also fail, etc., causing a progressive collapse of the entire strut system.

6.3.5 The design error

How could it happen that the designers overlooked this collapse mechanism? Unfortunately, they happened to use a wrong constitutive model, which overestimated the undrained shear strength of clay. For the Upper and Lower Marine clay layers in Figure 6.8, the average vertical effective stress are about $\sigma'_{v0} = 90$ kPa and $\sigma'_{v0} = 190$ kPa, respectively; the effective angle of internal friction $\varphi' = 22°$ and $\varphi' = 24°$, respectively. For these parameters, The Mohr – Coulomb model, used by the designers, estimates the undrained shear strength (Eqs. (6.11) – (6.12)) at

$$c_{u1} = 0.33 \times 90 = 30 \text{ kPa}, \quad c_{u2} = 0.35 \times 190 = 66 \text{ kPa}$$

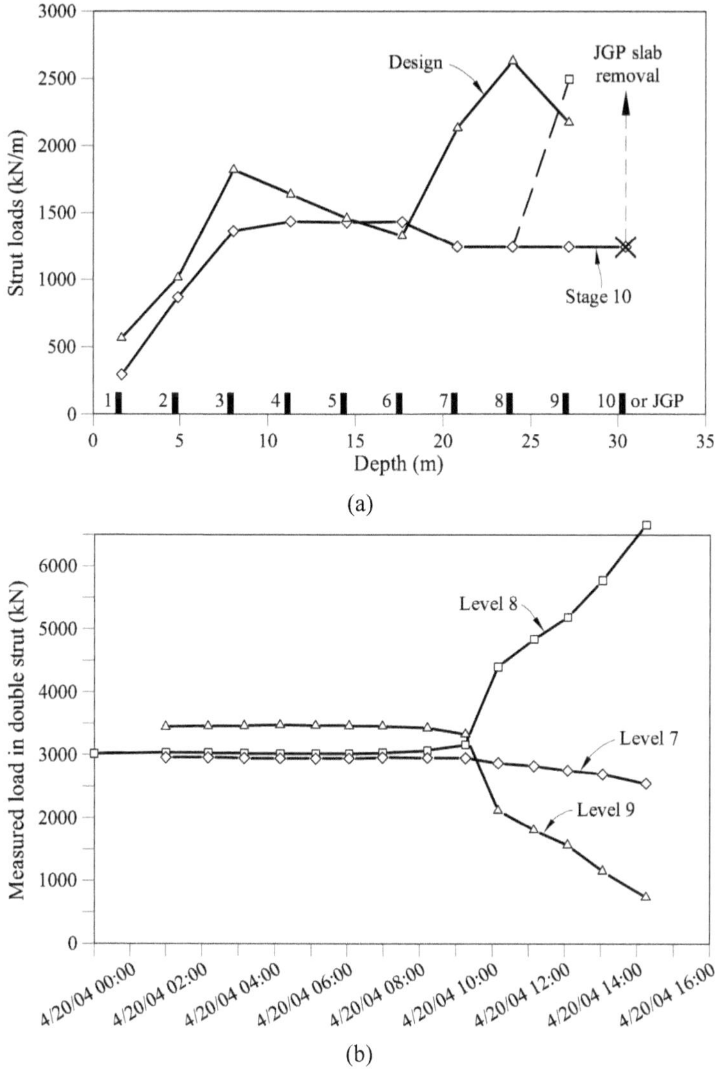

Figure 6.20 Redistribution of the strut loads after removal of the temporary JGP slab: (a) calculated; (b) measured in the strut S335 (Fig. 6.4) during the collapse between 10:00 – 15:00 on April 20, 2004 (after Davies *et al.*, 2006).

for the Upper and Lower Marine Clays, respectively. For comparison, the best estimate of the field and lab tests are (Fig. 6.8):

$$c_{u1} = 20 \text{ kPa}, \quad c_{u2} = 38 \text{ kPa}.$$

The total earth pressures for the short-term stability analysis are determined using Figure 6.16 with the corresponding strut loads at the end of the excavation calculated in Table 6.4, with the incorrectly calculated total active load being

$E_a = 10,090$ kN/m (when calculated correctly, the total active load is E_a = 11,717 kN/m). That is, the total active load was underestimated by almost 17%, and the load in the critical lower part of the excavation by almost 30%. Note that the designers used the factor of safety of 1.2 on the strut bearing capacity!

Table 6.4 Strut loads from the incorrect short-term stability analysis.

Strut loads (predicted) **0.25H=8.40**

Strut level	Layer	z (m)	Z (m)	y_i (m)	Y (m)	γ (kN/m³)	m (-)	c_u (kN/m²)	K_a (-)	$e_a = K \cdot \gamma \cdot D$ (kN/m²)	Strut load calc. FS=1 (kN/m)
0	UM	0.00									
	UM		1.60			16.00	0.80	30.00			
1	UM	1.60			3.20						266.24
	UM		3.20	3.20		16.00	0.80	30.00	0.81	166.40	
2	UM	4.80			3.20						798.72
	UM		3.20	6.40		16.00	0.80	30.00	0.81	332.80	
3	UM	8.00			3.20						1,181.44
	UM		3.20	9.60		16.00	0.80	30.00	0.81	416.00	
4	UM	11.20			3.20						1,331.20
	UM		3.20	12.80		16.00	0.80	30.00	0.81	416.00	
5	UM	14.40			3.20						1,331.20
	UM		3.20	16.00		16.00	0.80	30.00	0.81	416.00	
6	UM	17.60			3.20						1,331.20
	UM		3.20	19.20		16.00	0.80	30.00	0.81	416.00	
7	LM	20.80			3.20						962.56
	LM		3.20	22.40		16.00	0.80	66.00	0.59	300.80	
8	LM	24.00			3.20						962.56
	LM		3.20	25.60		16.00	0.80	66.00	0.59	300.80	
9	LM	27.20			3.20						962.56
	LM		3.20	28.80		16.00	0.80	66.00	0.59	300.80	
10	LM	30.40			3.20						962.56
	LM		3.20	32.00		16.00	0.80	66.00	0.59	300.80	
$Slab_B$	LM	33.60									
											10,090.24

In Figure 6.21, these incorrectly calculated strut loads are compared with the correct strut loads calculated using the best estimated undrained shear strength and with the actual design strut loads. As is seen, when the loads are underestimated, as was the case with the Mohr Coulomb model, removal of the JGP slab does not cause the failure of the ninth level struts. Indeed, even if the full load of the JCP slab is transferred to the ninth strut level, it leads to the strut load, which is 13% smaller than the design load. This is hardly surprising, because the design was based exactly on these underestimated strut loads and their redistribution during the removal of the JGP slab. Clearly, this underestimation of the earth pressure was one of the major reasons for the design error and excavation collapse.

6.3.6 Discussion

In spite of being significantly simplified, the above analysis confirms that overestimation of the undrained shear strength resulting in the underestimation of the active earth pressure, was one of the major reasons for the Nicoll Highway excavation collapse. It even managed to correctly predict the sequence of the events leading to the collapse: buckling of the walers in the sixth strut level and the critical effect of the removal of the JGP slab.

Clearly, it was an inappropriate use of the Mohr – Coulomb model in the FE analysis that brought this tragedy about. If the design was based, or at least compared to the above simplified analysis, this could urge the designers to review their FE calculations and use, e.g., one of the critical state models instead. The Nicoll Highway collapse could then, probably, have been prevented.

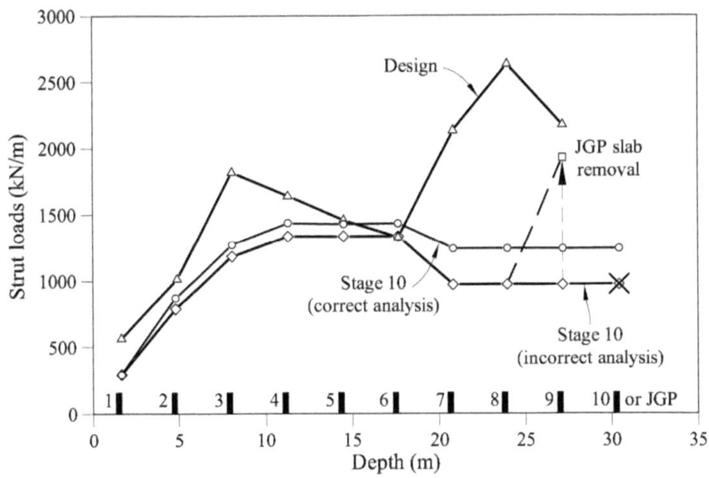

Figure 6.21 Strut loads from the incorrect short-term stability analysis.

6.4 Mitigation Measures

When the incident occurred, the surrounding ground collapsed into the excavation area (Fig. 6.22a). The area of the collapsed zone was approximately 100 by 130 m and 30 m deep. Its edge was only about 10 m away from the closest building – Golden Mile Complex.

The Nicoll Highway and the approach slab before the abutment of the Merdeka Bridge over the Kallang River were also damaged (Fig. 6.22b).

A storm drain located south of the cut and cover tunnel was the main drainage outlet, conveying water to the Kallang River. When the collapse occurred, the river water rushed into the collapse area.

Several key utilities, including power electric mains, gas mains, and water mains were broken by the collapse, causing a fire. All this represented a challenge for the following remedial works.

6.4.1 Immediate safety measures taken at the site

The damaged canal within the collapsed area was blocked up to prevent tidal water from the Kallang River from flowing into the site. Canvas sheets were placed on slopes around the site and any open areas to protect the soil and slopes. Cracks in the ground around the site were filled with grout to prevent water seepage. These efforts were made to prevent further movement within the collapsed area and ensure the safety of the rescue personnel and the adjacent structures.

(a)

(b)

Figure 6.22 The extent of the damage to: (a) the construction site; (b) the highway (After COI Report, 2005).

Surrounding buildings were inspected to ensure that they are safe for use. These buildings were continuously monitored and checked round the clock to ensure that they were stable and that the structural integrity of the buildings was not at risk. Additional settlement markers were installed in the nearest building, Golden Mile Complex, to monitor movements of the building edge closest to the collapsed site. Electrolevel beams (EL Beam) were installed to monitor movement of the building. Additional inclinometers were installed to monitor lateral deflections in the soil. Readings were taken every few hours. EL Beams and additional inclinometers were also installed at Golden Mile Tower (the next nearest building which is about 200 m from the incident site) to closely monitor this building.

The approach slab before the abutment of the Merdeka Bridge which is just next to the incident site, had also collapsed. A cut was made between the first and second spans of the bridge to separate the first span and mitigate the risk of the first span dragging down the entire bridge in the event of movement of the first span. Prism points (to measure *x-y-z* movements) and tiltmeters were installed to monitor the bridge.

6.4.2 Stages of recovery

The first stage of the recovery process was to start filling the immediate collapse site with "foam" concrete, a low strength and low viscosity concrete. This type of concrete can flow into the small voids between the debris and displace the water that has flowed into the excavation. This was to stabilize the void and prevent further soil movement and subsidence of the surrounding ground. The low strength of the concrete mix allows it to be removed later when excavation recommences following the recovery process.

The foam concrete was then topped by a layer of higher strength mass concrete in the second stage to form a roughly horizontal surface. The third stage was to carry out localized filling of mass concrete, or backfilling of soil, to stabilize the outer areas of the collapsed slopes. The fourth stage was to remove whatever debris (steel beams) that could be removed from the top, i.e. the mass concrete platform. The fifth and sixth stages were to backfill with soil below the Nicoll Highway and above the incident site up to original ground level. The seventh stage was to reinstate Nicoll Highway. The eighth and final stage of restoration was to backfill all other sunken areas up to original ground level.

After the above restoration process was completed, new retaining walls were installed to recommence excavation and resume construction.

6.4.3 Additional safety measures

The day after the incident, all contractors on the CCL were asked to carry out additional checks and further reviews, on the design and construction of all their temporary works. In addition, the owner and the project developer – Land Transportation Authority (LTA) – was asked to recheck in order to ensure that all the temporary works installed at site were done in accordance with the approved drawings. LTA also appointed an independent panel consisting of local geotechnical and structural experts to investigate the cause of the collapse. The panel had also been tasked to review the design and construction of the works at this site as well as other works on the Circle Line. LTA instructed that excavation works at all the other Circle Line sites were to be put on hold until the design of the temporary works and the constructed temporary works are thoroughly checked and found to be in order.

All the professional engineers who had designed the temporary works were been asked to confirm in writing to the regulatory agency, the Building Control Unit (BCU), that the temporary works designs would be adequate following their review. The temporary works designs have also been sent to the Building and Construction Authority (BCA) to conduct an independent design and site audit. The completion of the CCL1 stage was delayed until 2010 (by four years).

The Government has also appointed a Committee of Inquiry (COI) to ascertain the causes and circumstances of the incident and make recommendations to prevent the recurrence of such an accident in future.

6.5 Lessons Learned

The final report of the Committee of Inquiry into the collapse of the Nicoll Highway was submitted on May 11, 2005. Faced with the monumental task of piecing together information from 193 witnesses and volumes of documents, the COI elucidated the crucial errors and persons responsible for the collapse of Nicoll Highway.

In the COI's words: "Warnings of the approaching collapse were present from an early stage but these were not taken seriously". It concludes that "the death of four persons was the direct result of the collapse" and that "the Nicoll Highway collapse could have been prevented". The following recommendations were made by the Committee of Inquiry to avoid similar incidents in the future.

6.5.1 Effective risk management

The potential for major accidents, whether due to the construction process or deficiencies in design, must be recognized and expeditiously controlled. It is inappropriate to leave the control of risk wholly to contractors. Owner's and builder's management must seek a balance between production pressures and quality and safety goals.

6.5.2 Robustness of design

A robust design is essential. This robustness is provided by identifying the hazards and checking that the proposed design can adequately withstand them. The design should have sufficient redundancy to prevent a catastrophic collapse in the event of a failure of any particular element. Temporary works for deep excavation should be given the same respect as permanent works.

6.5.3 Numerical modeling in geotechnical design

Numerical analysis should supplement and not supplant sound engineering judgment and practice. Those who perform geotechnical numerical analysis must

Chapter 7

Tunnel Excavation Collapse: Borràs Square, Spain

TABLE OF CONTENTS

A.M. Puzrin et al., *Geomechanics of Failures*, DOI 10.1007/978-90-481-3531-8_7,

Chapter 7

Tunnel Excavation Collapse: Borràs Square, Spain

7.1 Case Description

In 1991, the city of Barcelona was actively building a new transportation infrastructure to prepare for the Olympic Games of August 1992. An important utility was a 35 km long freeway ring (the "Rondas") which substantially improved the city traffic in the years to come. The failure described here refers to an auxiliary two-way underground branch which was built to connect a radial highway leaving Barcelona from the north of the city and the Ronda ring. Figure 7.1 shows the crossing of the two main highways (Via Augusta and Ronda de Dalt) and the so-called I-J branch (dashed lines). The failed section is also indicated in the figure.

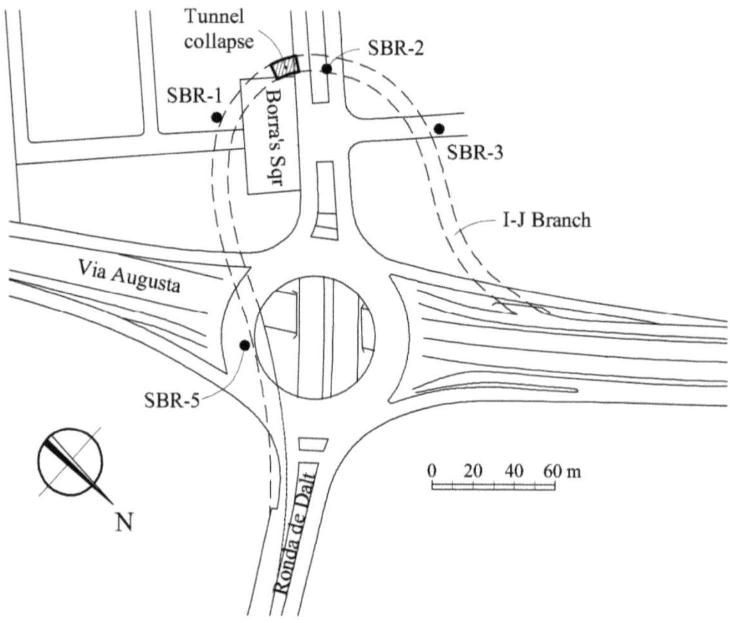

Figure 7.1 Plan view of tunnel showing the failed section and the position of borings.

The soil was investigated by a few borings (SBR-1,-2,-3 and -5) whose position is also shown in Figure 7.1. The tunnel was being excavated in the direction SBR-5, SBR-1 to SBR-2.

Most of Barcelona Metropolitan Area is founded on ancient quaternary soils, which are described as a sequence of reddish clays and yellowish silts with variable proportions of gravel and some carbonate crusts. These are stiff deposits whose thickness decreases towards the north in the direction of the Collserola mountain range which acts as a natural limit for the city. The tunnel is located in the northern part of the city, where the quaternary deposits, progressively thinner, overlay the rock substratum (decomposed granite, shales and metamorphic rocks). The longitudinal profile of the entire I-J tunnel, as interpreted from the borings, is shown in Figure 7.2.

The tunnel failure is located between borings SBR-1 and -2. The profile shows the small tunnel cover (5.5 m in the failed section). The tunnel was entirely excavated in quaternary soils, although the granite substratum was close to the invert. Figure 7.2 shows the difficulty in establishing a clear layering pattern. The soil at the location of SBR-2 was described as a sequence of clayey gravels, carbonated clays, clayey gravels again, clayey sands, and finally the granitic substratum. Also relevant is the presence of some "fills" difficult to identify because they were also described as clay and gravel. The profile in Figure 7.2 indicates also recorded SPT values. They are typical of medium to hard soils. The ancient quaternary deposits in Barcelona are in general unsaturated.

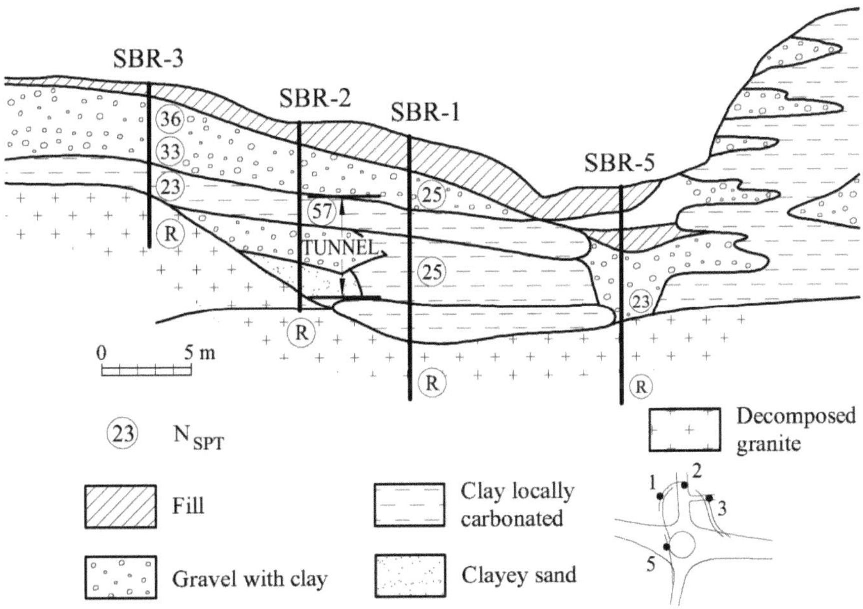

Figure 7.2 Longitudinal geotechnical profile along the tunnel.

7.1.1 Construction procedure

The tunnel was built manually in a classical head and bench excavation. Figure 7.3 shows a cross-section, the dimensions of the head and bench, and the tunnel cover. The head was excavated ahead of the bench (12 m approximately) and was excavated in full, in unsupported lengths of 1.08 m and it was immediately protected by heavy double T, wide wing, (HEB 160) steel sets. Perforated steel "Bernold" plates, supported by the sets, completed the steel reinforcement of the vault. Concrete was pumped between the Bernold plates and the excavated soil surface to form a continuous monolithic lining, 25 cm thick. Wire mesh and a shotcrete layer, 6 cm thick, were finally applied. Figure 7.3 shows a cross section of the installed lining support and Figure 7.4 a schematic detail of a horizontal cross-section of the tunnel lining. Daily advance was 2.16 – 3.24 m (two to three steel sets installed).

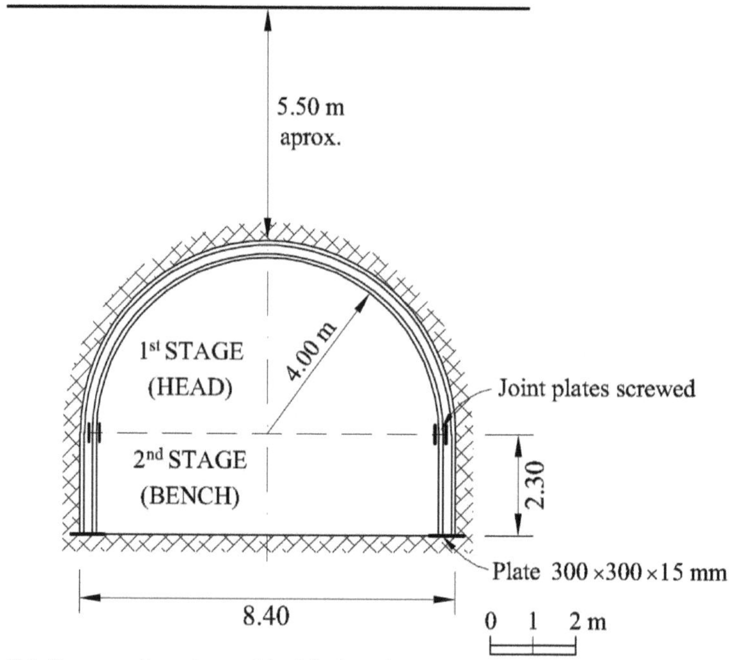

Figure 7.3 Cross-section of tunnel in failed section.

The second construction phase involved the excavation of the bench. In cross-section it had a width of 8.40 m and a height of 2.30 m. A critical operation of this phase was the appropriate underpinning of the vault. To do so, the sets were lengthened by means of straight and vertical HEB 160 poles. The same support (Bernold plates, poured mass concrete, and a final shotcrete layer) was also applied. Figure 7.5 shows the side walls of the tunnel in construction with the exposed Bernold plates between HEB poles, before the application of the shotcrete layer. Bench excavation started by opening a central passage, 4 m wide at the base (Fig. 7.6).

The vault was thereafter supported by two longitudinal soil buttresses with the approximate dimensions indicated in Figure 7.6. Right and left buttresses, however, had an offset distance which ranged between 4 and 12 m. Specifications required that the vault should be underpinned by excavating the buttress in widths not exceeding 2.16 m (two new vertical poles added).

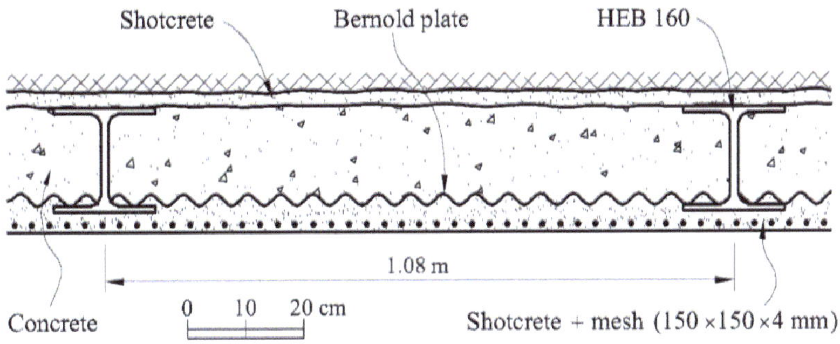

Figure 7.4 Horizontal cross-section of tunnel lining.

Figure 7.5 Perforated Bernold plates, placed between steel sets, act as a "skin" reinforcement.

The bench advance rate was faster than the head excavation and daily figures ranged between 3.24 and 6.48 m. The preceding description corresponds to project specifications which were essentially followed during construction.

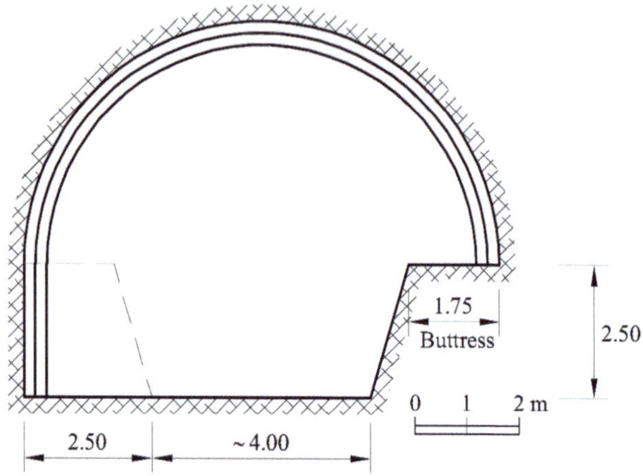

Figure 7.6 Bench excavation procedure.

Figure 7.7 The finished tunnel.

7.1.2 The failure

The collapse destroyed a length of 11 m of tunnel which was being finalized ahead of a fully supported section. Figure 7.7 is a picture of the completed tunnel. The collapsed section is located at the opposite extreme of the photograph, which was taken a few days after the failure. Figure 7.8 is a sketch of the construction situation of the 11 m long tunnel section prior to failure. The left side of the tunnel was almost complete. The vault had already been underpinned on the left side. A worker, located on the left of the tunnel, was welding steel bars between successive steel poles to improve their lateral stability. The right buttress was in place, supporting the semicircular head steel sets.

On the morning of June 26, 1991 a team of workers began to excavate the right buttress in order to underpin the vault. It seems that they started work on the far right corner of the tunnel, just touching the vertical front. A backhoe loader and a wheel front loader were moved into the working area to assist the team. Failure occurred immediately after the workers had excavated two or three meters of the right buttress. They later declared that the vault collapsed on them. They managed to escape, protected by the machines. However, the welder working on the stabilization of the steel poles on the left side of the tunnel was killed by the roof collapse. The tunnel failure left a 12 m high "sinkhole" or cavern of approximately cylindrical shape (Fig. 7.9).

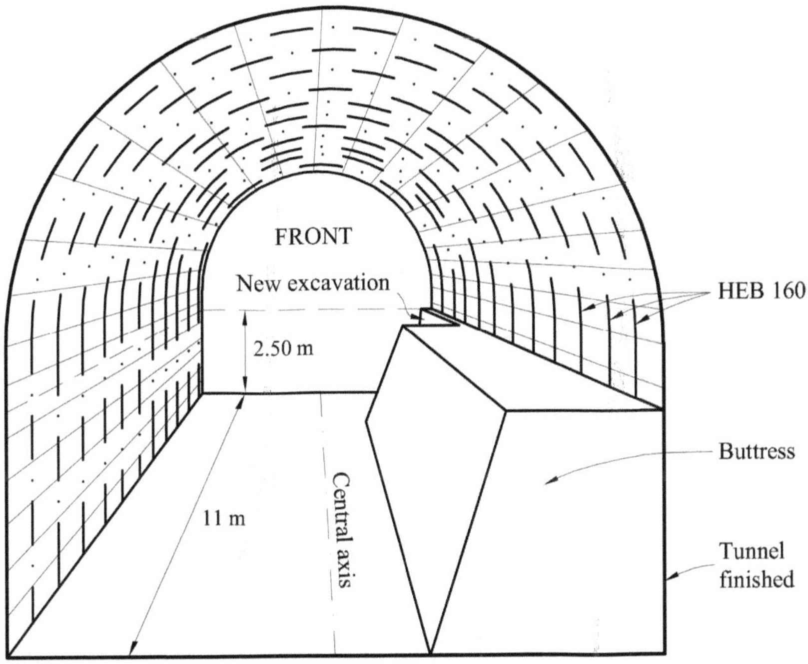

Figure 7.8 Final excavation operations before failure.

A plan view on the failure is given in Figure 7.10. The approximate diameter of the hole at the street level was 7 m (the dimensions of the base of the failed tunnel section were 8×11 m).

Some interesting observations were given in a report issued by the fire department of Barcelona (the firemen were involved in rescue tasks immediately after the failure). Initially, the collapse though did not reach the street level. It took several hours before this happened. The collapsed vault was only "loaded" by a relatively thin layer of soil whose thickness was estimated visually as being around 1 m. Further observations after the second and definitive collapse, once the debris and machines were removed, provided additional details (see Figs. 7.9 and 7.11).

Figure 7.9 shows the finished tunnel as it could be observed from the open cavern. The exposed soils had a significant clay and silt content. The picture also shows that overbreaks occurred during vault excavation. They were filled by concrete.

When the collapsed section was excavated, buried steel sets, which once supported the vault, were uncovered (Fig. 7.11). The sets and the vertical steel supports had tilted towards the right buttress, as shown in the photograph. These observations indicated that the failure was most probably preceded by a loss of support of the sets resting on the right buttress. In other words, the instability of the right soil buttress, loaded by the vault, was at the origin of the catastrophic failure.

Figure 7.9 View of the cavern left by the failure. Note the size of the fireman on the lower part of the photograph.

In order to analyze this case, we will proceed as follows: First, the stability conditions of the cavern will be formulated with the aim of deriving average strength parameters for the soil. Then the actual soil load against the vault, knowing that the right support was yielding, will be investigated by considering arching effects in the soil. Finally, the failure conditions of the right soil buttress, loaded by the vault, will be examined.

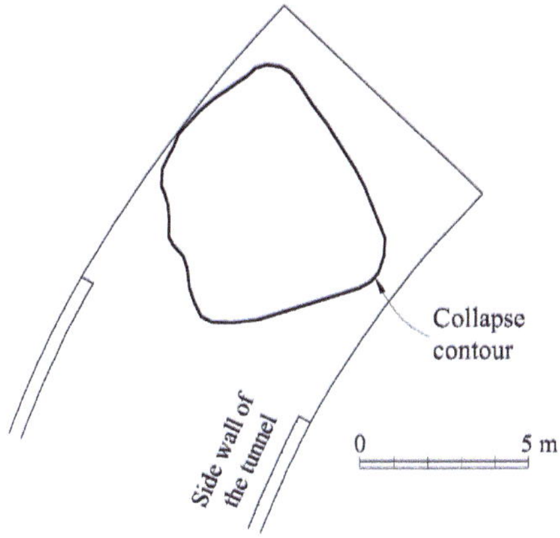

Figure 7.10 Plan view of the failure.

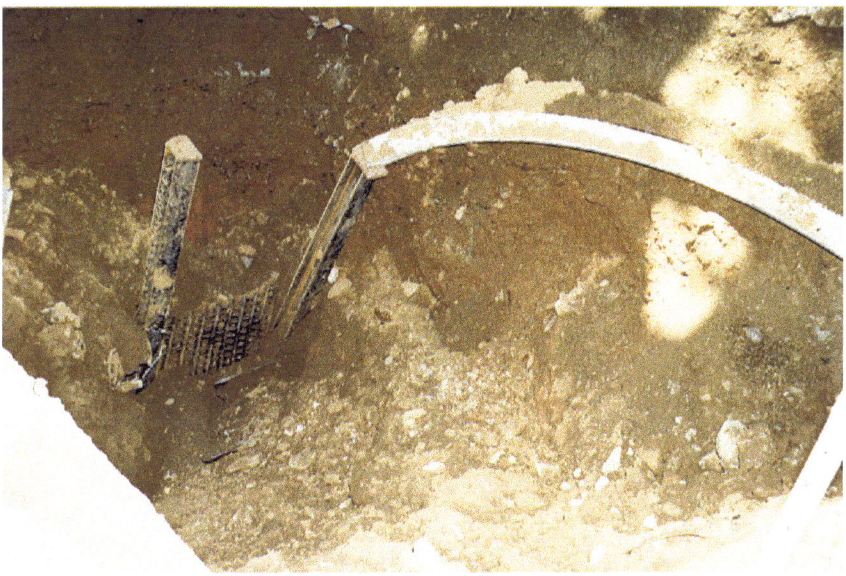

Figure 7.11 Steel sets and poles, tilted sets towards the right buttress after the failure.

7.2 Back Analysis of Stability Conditions of the Cavern

The failed tunnel section, 8.40 m wide and 11 m long will be approximated by a half sphere, 5 m in radius *R*, whose center is located 10.5 m below the ground surface (Fig. 7.12). The tunnel collapse created a cavern reaching the surface, having an approximate truncated conical shape. The upper contour of this truncated cone is shown in Figure 7.10. Conditions leading to the instability of this mass of soil on top of the failed tunnel will be used to characterize the soil strength, which could be approximated by a (c,ϕ) Mohr – Coulomb failure criterion. Since the soil is unsaturated, no distinction between effective and total strength parameters will be made.

A drained mechanism will be assumed and stability conditions will be solved by means of the upper bound theorem of plasticity. Therefore, in order to comply with the associativity rule of the plastic theorem (the dilatancy angle on internal failures surfaces is given by the friction angle) and to ensure a kinematically admissible mechanism, the mobilized failure cone must have a vertex angle of 2ϕ. This is shown in Figure 7.12. The motion of the cone is vertical. The cone-sphere interaction defines a spherical "dome" which limits the cone in its lower part. The radius of the circle of intersection (r_1 in Fig. 7.12) will be controlled by angle α. In fact, the geometry of the family of cones considered in the optimization process will be characterized by the angle α. Note that, given a depth of the failed tunnel section (sphere), the apex of the cone may be located inside the soil or above the upper surface depending on the friction angle.

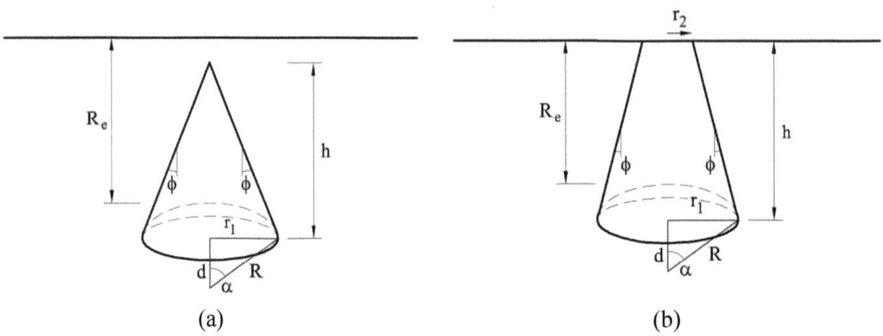

Figure 7.12 Geometry of the cavern failure: (a) cone vertex remains inside the soil; (b) cone vertex above the ground surface.

The geometry in Figure 7.12 is defined by the following variables:

$$d = R\cos\alpha, \tag{7.1 a}$$

$$r_1 = R\sin\alpha, \tag{7.1 b}$$

$$h_{max} = R_e + R - d, \tag{7.1 c}$$

$$h = \min(h_{max}, r_1/\tan\phi), \tag{7.1 d}$$

$$r_2 = \max(0, r_1 - h\tan\phi), \tag{7.1 e}$$

where R is the radius of the sphere ($\approx$5 m), R_e is the tunnel cover ($\approx$ 5.50 m), h is the height of the cone apex above the base circle of the cone (when the apex remains inside the soil), h_{max} defines the height of the truncated conical volume (cone apex above the soil surface), r_2 is the radius of the upper face of the truncated cone in the second case, and d is an auxiliary variable. The volume (V) and outer lateral surface (A) of the cone are given by

$$V = \frac{1}{3}h\pi(r_1^2 + r_2^2 + r_1 r_2) - \frac{1}{3}\pi(R-d)^2(3R-(R-d)), \tag{7.2 a}$$

$$A = \pi(r_1 + r_2)\frac{h}{\cos\phi}. \tag{7.2 b}$$

Under a virtual vertical displacement rate, δ, the upper bound theorem, requires the balance of the rate of work performed by the external forces (cone of soil having a unit weight γ) and the internal dissipation of work on the failure surface. The former is given by

$$W^{\text{ext}} = V\gamma\delta. \tag{7.3}$$

In a drained shearing, the energy dissipation per unit area is given by $c \cdot \delta_t$, where c is the cohesion and δ_t is the displacement component in the direction of the failure surface (see Atkinson, 1981). Since $\delta_t = \delta\cos\phi$ and

$$W^{\text{int}} = cA\delta\cos\phi, \tag{7.4}$$

the condition $W^{\text{ext}} = W^{\text{int}}$ leads to an explicit expression for c:

$$c = \frac{V\gamma}{A\cos\phi}, \tag{7.5}$$

where V and A are functions of α (Eq. 7.2).

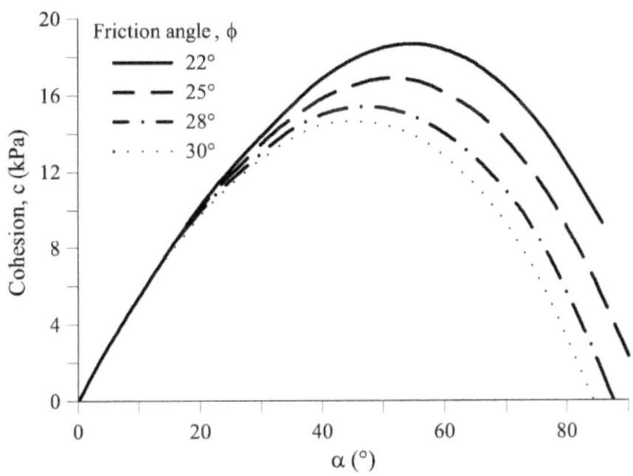

Figure 7.13 Values of cohesion c derived from the application of the upper bound theorem of plasticity.

For a given value of ϕ, Equation (7.5) provides values of c, which depend on the geometry (defined by α) that led to collapse. We are looking for the maximum value of c (in terms of α) required for the equilibrium in order to find the upper bound that, hopefully, will be close to the true cohesion.

Finding the maximum of $c(\alpha)$ in (7.5), in closed form, is a complex task. But it is a simple matter to find it by means of an Excel solver. The calculated values of c for varying α, for four values of ϕ are given in Figure 7.13. The selected range of ϕ values, $22° - 30°$, cover the expected friction angles of the excavated soil. The calculated cohesion values ($14 - 18$ kPa) are relatively low, but consistent with the instabilities (overbreaks) suffered during head excavation.

For $\phi = 28°$, a reasonable value, the critical value of α is around 50° (see Fig. 7.13). The vertical cohesion leading to strict equilibrium is $c = 15.4$ kPa. Then, Equations (7.1) predict a critical failure cone having a vertex inside the ground but close to the surface.

The actual failure progressed upwards and daylighted at the Borràs square, probably because the soil strength worsened close to the soil surface.

Let us now consider the problem of determining the soil load on the tunnel vault.

7.3 Arching Effects and Loads on the Tunnel Vault

7.3.1 Theory

In his book on theoretical soil mechanics, Terzaghi (1943) developed an arching theory which may be used to estimate the loads actually exerted by the soil on the tunnel vault. The classical concept involves the yielding of part of the lower horizontal support of a mass of soil. The downward motion of the soil is opposed by shear forces developed in a surface (ab, dc) extending from the lower yielding base towards the upper soil surface (Fig. 7.14).

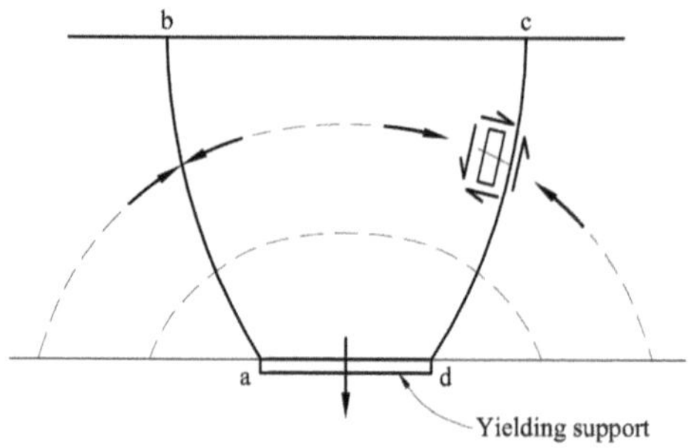

Figure 7.14 Arching effects induced by a yielding support.

In our case, the yielding support is the tunnel vault. When the vault settles, the soil above will be partially supported by internal shearing. This load transfer mechanism reduces the vertical load applied to the vault. If principal stresses are considered (Fig. 7.14), major principal stresses tend to form "discharge" arches in the vicinity of the yielding support. These internal arches are also a convenient mechanical interpretation of the phenomenon and help to explain the increase in normal stress on both sides of the yielding support.

Consider in Figure 7.15a a yielding support of circular shape and radius R on the bottom of a soil layer of thickness H. For simplicity, the resisting shearing will be assumed to act on a cylindrical surface of radius R. Consider the vertical equilibrium of a disk of thickness dz, located at a depth z. Following the notation given in Figure 7.15, the following relation holds:

$$\left(\sigma_v + \mathrm{d}\sigma_v\right)\pi R^2 + 2\pi R K \sigma_v \tan\phi \mathrm{d}z + 2\pi R c \mathrm{d}z = \pi R^2 \sigma_v + \pi R^2 \gamma \mathrm{d}z, \quad (7.6)$$

where γ is the soil specific weight and K is the coefficient of earth pressure, assumed to be constant.

The solution of the differential Equation (7.6) for the vertical stress σ_v, for the condition of no vertical stress applied on the surface, is

$$\sigma_v = \frac{R(\gamma - 2c/R)}{2K \tan\phi}\left[1 - \exp\left(-\frac{2Kz}{R}\tan\phi\right)\right]. \quad (7.7)$$

Equation (7.7) has been represented in Figure 7.16 for three values of the at-rest pressure coefficient (K = 0.8, 1 and 1.2) and for a set of parameters which characterize the arching conditions developing above the yielding vault (R = 5 m; γ = 20 kN/m^3, c = 15.4 kPa, ϕ = 28°). The strength parameters were back calculated in the previous section from the global collapse observed a few hours after the initial failure of the vault took place.

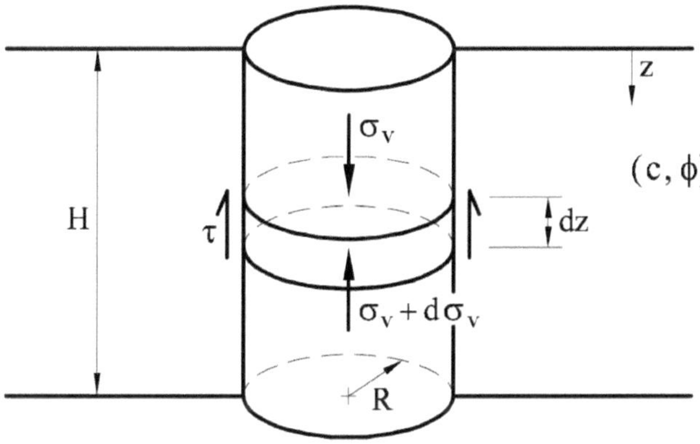

Figure 7.15 Mobilized cylindrical soil body above a circular yielding base. Arching effects above the yielding vault.

Terzaghi (1943) reported that observations in sands indicated that the value of K increased from about one immediately above the center line of the yielding strip (he considered a two-dimensional problem) to a maximum of about 1.5 at an elevation of $2R$ above the strip. Therefore, a value K = 1.2 seems reasonable in our case.

The solution for K = 1.2 yields a vertical stress of 42.6 kPa for a depth below the surface of around 6 m. It is not reasonable to pinpoint stress values at larger depths because the presence of the vault implies a weight defect (if compared with the full cylinder of the theoretical solution), which would lead to smaller σ_v values than those represented in Figure 7.16.

Terzaghi (1943) also mentioned that if the depth of the yielding support (z) increases, the yielding of the lower support is not noticed in the upper layers. He specified that arching effects disappear at heights in excess of $5R$ above the yielding base. As a result, the proposed distribution of vertical stresses resulting from Terzaghi's analysis is shown in Figure 7.17. The soil develops the full geostatic stress profile above the 5*R* limit mentioned and then the stress decreases until it eventually joins the prediction of Equation (7.7).

In the case analyzed here, the distance between the yielding vault and the soil surface is about 1.1*R* and therefore full arching is expected to develop. This justifies the use of Equation (7.7) to find the vertical stresses at the level of the vault.

The resisting mechanism of the failed portion of the tunnel is complicated, structurally speaking, because the vault was continuous across the limit between the tunnel already finished ("stable" tunnel in Fig. 7.18) and the failed length. It is reasonable to accept that during yielding of the failed section, some of the soil load exerted against the vault, even if it is reduced by arching effects, would be shared by the stable tunnel. Vaults are known to develop strong three-dimensional interactions against deformation.

It is also true that no reinforcing was acting in the longitudinal tunnel direction (a common situation in tunneling construction).

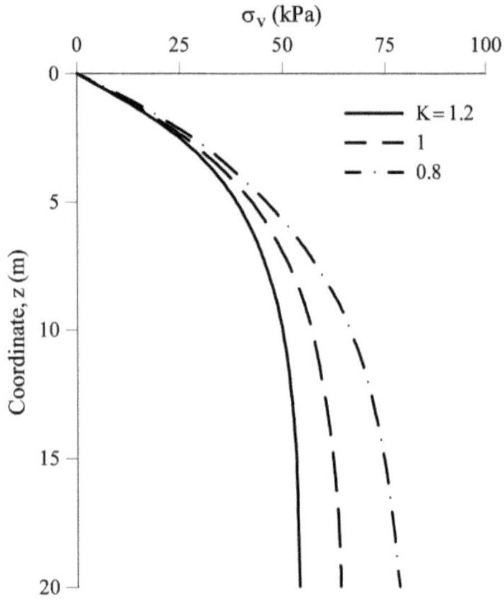

Figure 7.16 Vertical stresses over a circular yielding base.

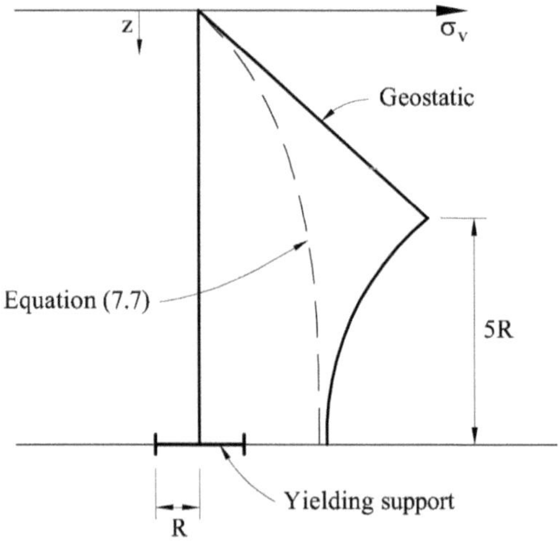

Figure 7.17 Vertical stresses above a yielding support, following Terzaghi (1943).

If half of the (average) vertical stress was transferred to the stable tunnel the vault would have received only a small "pressure" estimated in 42.6 kPa/2 = 21.3 kPa which roughly corresponds to observations of the firemen in the sense that a layer of soil whose thickness was estimated in 1 m accompanied the vault during its fall.

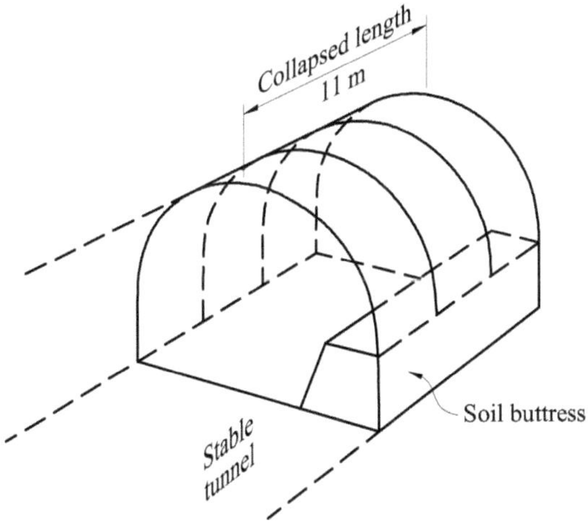

Figure 7.18 Stable tunnel and failed section. Vault was continuous throughout.

7.4 Buttress Failure

The foundation of the vault on top of the buttress is schematically indicated in Figure 7.19. The vault, subjected to a vertical stress σ_v, is resting on the sloping buttress which is about to be excavated in order to underpin the steel sets. The stability of this slope, loaded on top, may be analyzed by different procedures. A very simple approach is to examine a planar slide (sliding surface AC, inclined by angle β) loaded on top by a line loading T.

The force T may be derived by a simple equilibrium in vertical direction applied to a quarter of the tube:

$$T = \sigma_v R. \tag{7.8}$$

The equilibrium of wedge ABC in vertical and horizontal directions leads to

$$\begin{aligned} T + W &= N\cos\beta + S\sin\beta, \\ N\sin\beta &= S\cos\beta. \end{aligned} \tag{7.9}$$

The shear force S is expressed in terms of the mobilized cohesion ($c^{\text{mob}} = c/SF$) and friction ($\tan\phi^{\text{mob}} = \tan\phi/SF$) (where SF is the safety factor):

$$S = N\tan\phi^{\text{mob}} + c^{\text{mob}} L. \tag{7.10}$$

where L is the length AC (Fig. 7.19).

The solution of (7.9) and (7.10) leads to the following expression for the safety factor:

$$SF = \frac{N \tan\phi^{mob} + c^{mob} L}{N \tan\beta}, \qquad (7.11)$$

where

$$N = \frac{T + W}{\cos\beta + \tan\beta\sin\beta}. \qquad (7.12)$$

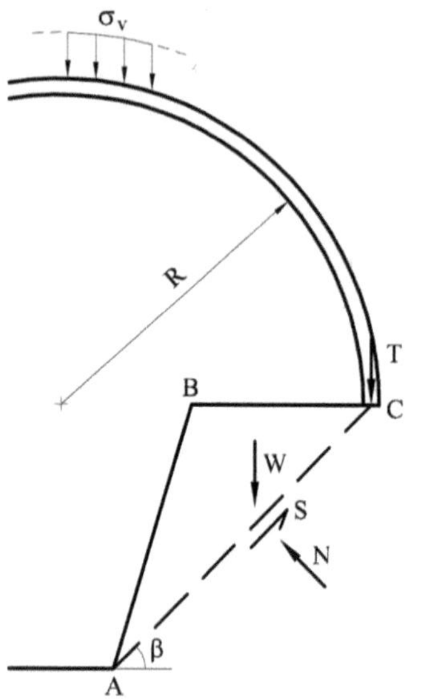

Figure 7.19 Stability conditions of right buttress loaded by vault arch.

Equation (7.11) has been plotted in Figure 7.20 for increasing values of *T*, for the buttress dimensions left after the excavation of the lower bench section of the tunnel (Fig. 7.6), and for a soil unit weight of 20 kN/m^3. The selected strength parameters correspond to the pair ϕ = 28°, c = 15.4 kPa, which explains the full collapse of the tunnel analyzed in Section 7.2.

The actual load per unit length acting on the top of the buttress may be approximated by Equation (7.8). This equation implies an infinite extent of the vault in the longitudinal direction. However, the failed length, 11 m long, had a common vault with the remaining part of the tunnel already built. In other words, the continuity of the vault provided an additional support which is difficult to take into account without a specific structural analysis, three-dimensional in nature (see Fig. 7.18).

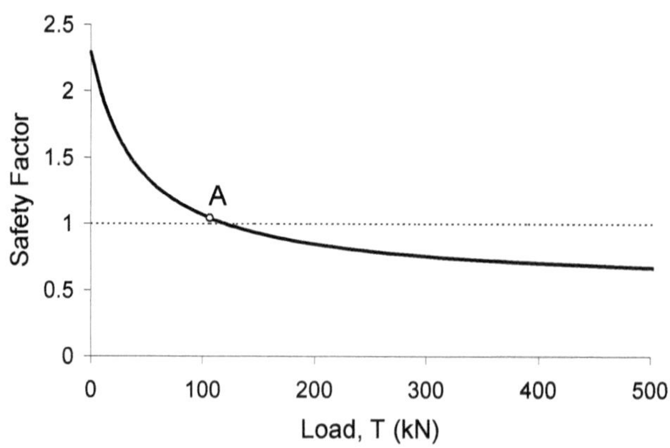

Figure 7.20 Safety factor in terms of the load *T* transmitted by the vault to the top of the buttress. Point A indicates the state before starting the work in the morning of June 26, 1991.

Vaults are very effective supporting structures but in our case it did not have any longitudinal steel reinforcement. It has roughly been estimated that the contribution of the remaining of the tunnel to support loads on the final section, which was close to completion, amounts to half of the loads provided by the formula (7.8). In this way, the calculated average load on the soil buttress is $T = 42.6\ \text{kPa} \times 5\ \text{m} \times 1\text{m}/2 = 106.5\ \text{kN}$, which is shown in Figure 7.20 as point A.

Figure 7.20 indicates that, at the time of starting the work in the morning of June 26, 1991, the safety factor against failure of the soil buttress was low (≈ 1.05). Note also that, without any "vault effect", failure could have taken place at an earlier date.

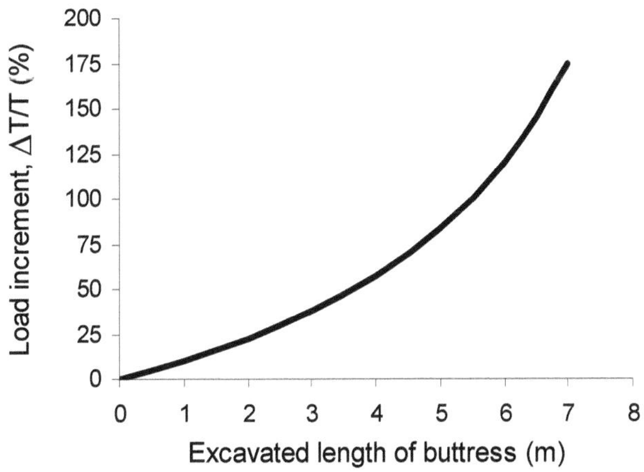

Figure 7.21 Increment of load *T* transmitted by the vault to the buttress when a niche of increasing length is excavated.

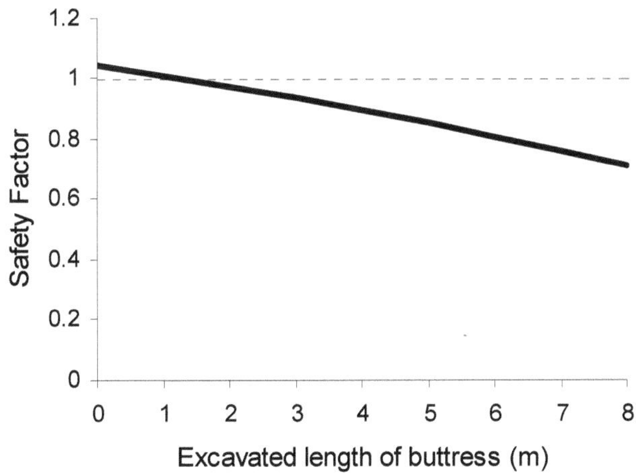

Figure 7.22 Calculated reduction in safety factor as the buttress excavation progresses.

Now consider the effect of opening a lateral niche to underpin the steel sets. The length of the bench (initially 11 m) would be progressively reduced. If a simple calculation is made (the average line load exerted by the vault is linearly dependent on the buttress length actually supporting the vault) the following expression is calculated for increment of load ΔT over the initial value (before any underpinning excavation), T :

$$\Delta T / T = e_L / (11 - e_L) \tag{7.13}$$

where e_L is the excavated length (in m). This equation is plotted in Figure 7.21.

Excavation lengths of 2, 3 and 4 m result in load increments of 22, 37 and 57% over the initial value. This load increment implies a reduction in safety factor which has been indicated in Figure 7.20. For an excavated length of 1.5 – 2 m the safety factor is already 1.0 and failure is imminent. The evolution of safety factor has also been represented in Figure 7.22 against the excavated length of the bench.

7.5 Discussion

Three different techniques, applied in a proper order, have been used in this chapter to analyze a complex problem of a marked three-dimensional nature. The development of a large cavern day-lighting at the surface was the starting point of the analysis. The soil strength which explains this failure was used in the subsequent examination into the reasons for the failure. Since the soil involved was unsaturated, a drained analysis was performed. The upper bound theorem of plasticity offered a convenient tool to handle the approximately conical shape of the failure. The back analysis provides only a combination of c and $\tan\phi$ parameters, leading to failure. One of them (typically the friction angle, which could be correlated with soil plasticity and granular content) has a lower uncertainty and may be more easily approximated in a range appropriate for clayey soils ($22° - 30°$). However, this difficulty is more apparent than real because alternative (c, $\tan\phi$) pairs provide similar results in subsequent calculations.

It is felt that the back-analysis of the final collapse provided a good approximation for the strength parameters operating on the soil surrounding the tunnel.

The analysis of the failure of the right buttress indicated that failure conditions were achieved for a relatively minor proportion of the expected geostatic load against the vault. This has been attributed to arching effects. The case shows that, even for shallow tunnels, arching may be very effective reducing the actual surcharges. The classic Terzaghi (1943) theory of soil arching, applied to a yielding cylinder, has proved useful to approximate arching in this case. Recent work by Potts and Zdravkovic (2008), using elasto-plastic finite elements, supports the analysis of Terzaghi. Of course, some simplifications need to be done in order to accommodate the real geometry to an amenable case but the reduction in apparent loading attributed to arching is consistent with the final step of the analysis, namely the instability of the "right" buttress under vault loading. A simple calculation procedure (limit equilibrium of a loaded wedge) explains the failure. More sophisticated calculation procedures are probably of doubtful relevance in this case, because the key aspect of the entire analysis is the correct selection of the strength parameters (first stage) and the arching effects (second stage).

7.6 Mitigation Measures

The tunnel collapse was a consequence of an inappropriate construction method designed for rocks, even if they exhibit a low strength. An excavation procedure suited for soil would have prevented the failure. Among such procedures, a popular and classic construction method is the so-called "Belgian method" which requires a relatively firm soil. The Belgian or "flying arch method" is described in tunnel texts and handbooks (Széchy, 1967; Bichel and Kuesel, 1982). In this method (Fig. 7.23) excavation starts as a small central heading (approximately 1 m wide × 2 m high) which is supported by poles and timber planks (Fig. 7.23a). This initial opening is widened on both sides in order to create a vault (Fig. 7.23b). The support is provided by ribs supported on frontal poles, extending from a central unexcavated mass of soil, and the concrete arch already built (Fig. 7.23c). Timber or steel planks driven by a vibrating hammer support the exposed soil between the ribs. Then an arch lining, supporting the vault, is concreted (Fig. 7.23d). A central unexcavated soil buttress is maintained to stabilize the front. The vault arch is then underpinned by relatively narrow excavations or pits which allow the construction of side walls in order to transfer the vault load towards the level of the invert (Fig. 7.23e,f). The process ends when the entire section is excavated and a curved invert (if necessary) is laid down (Fig. 7.23g). The entire process is labour-intensive and requires specialized workers. However, the method, which has variants when moving from country to country or even from city to city, has been widely used when the length of the tunnel or other specific situations prevents the use of modern shield tunnelling.

Having in mind the reasons for the failure, it is also quite possible that the method used could have been successful, provided the bench excavation procedure was modified. In fact, starting at the situation of a "long" central corridor, leaving soil benches on both sides, a more careful underpinning operation excavating lateral openings of a smaller length (say 1 m) could have prevented the failure. An even better procedure is to advance the excavation of the lower half of the tunnel at a slower rate, avoiding the long central corridor. This would enhance three-dimensional effects and help maintaining the stability of the vault during the installation of HEB poles and the construction of side walls.

7.7 Lessons Learned

7.7.1 Bench excavation: a frequent source of tunnel failures

A successful excavation of the tunnel head and the formation of the protective vault is not an indication of a job "almost done". The excavation of the bench has to be approached from a structural perspective in the sense that tunnel operations are in fact an underpinning task. In the analyzed case, there was probably a limited understanding of the stability conditions of the excavated bench. It should be added that the head and bench tunnelling procedure selected, which is well suited for soft rock materials, is at the limit of its capabilities when dealing with soils of medium strength.

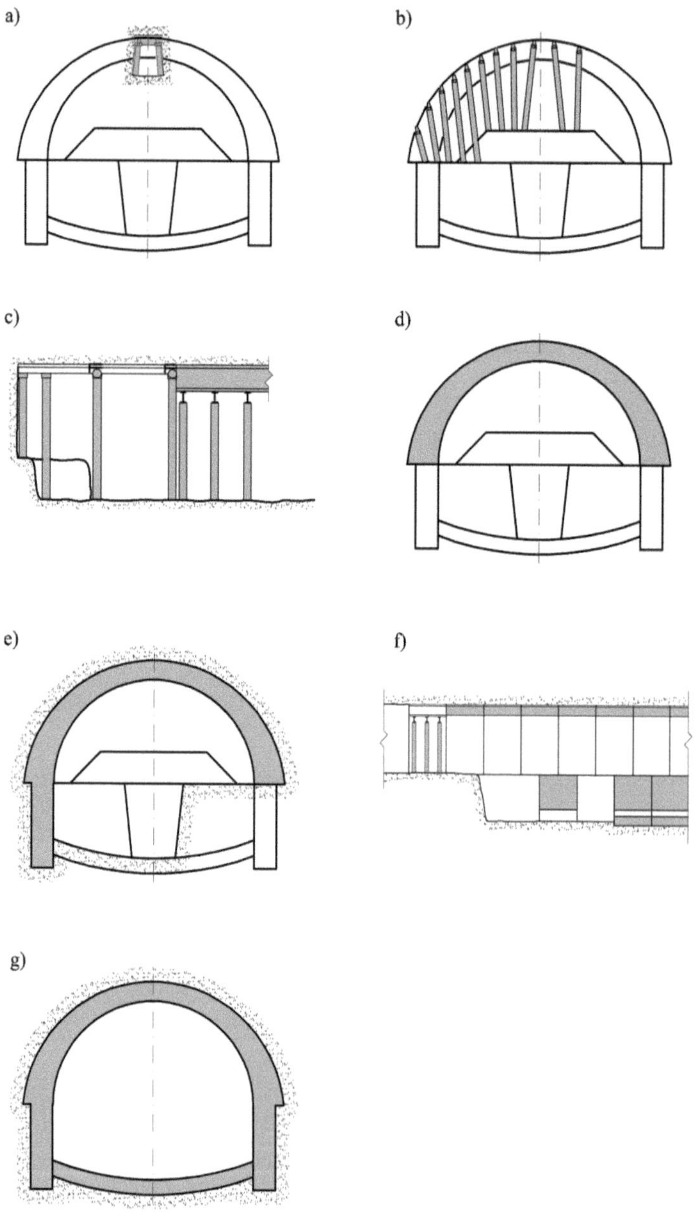

Figure 7.23 Belgian method of tunnel construction. Explanation given in the text.

7.7.2 The benefits of arching

A designer of the tunnel, being aware of the need to ensure stability of the foundation of the vault during bench excavation, would not dare reducing the overburden stress below its nominal value (geostatic stresses). The reason is the low cover to diameter ratio. The case indicates that, even for these shallow tunnels, arching helps to increase safety factors. It was not capable, however, of ensuring adequate safety in this case, because other design and construction specifications (namely, the length, cross-section of the right buttress, and underpinning operations) reduced the safety.

7.7.3 Strength parameters from back analysis of failures

Back analyses of stability failures provide good estimations of field strength. Pore water pressures are typically a difficulty in drained calculations because they are often unknown at the time of failure. In the case described the soil was not saturated and the analysis was performed in total stresses. No attempts were made to directly introduce suction effects because of a lack of data.

7.7.4 Structural strength of support is generally not an issue in shallow tunneling

Bending moments, normal and shear forces in steel sets were far from reaching critical conditions in this case. This check (not included here) was necessary to discard a structural tunnel failure. This is generally the case in shallow tunnels. The risks in shallow tunneling are associated with soil instability.

References

Atkinson, J.H. (1981) *Foundation and Slopes. An Introduction to Applications of Critical State Soil Mechanics*. McGraw Hill, New York.

Bickel, J.O. and Kuesel, T.R. eds. (1982) *Tunnel Engineering Handbook*. Van Nostrand Reinhold Co, New York.

Potts, V.J. and Zdravkovic, L. (2008) Finite element analysis of arching behaviour in soil. *Proceedings of the 12th International Conference of International Association of Computer Methods and Advances in Geomechanics* (IACMAG). Goa, India. 3642 – 3649.

Széchy, K. (1967) *The Art of Tunneling*. Akadémia Kiadó, Budapest.

Terzaghi, K. (1943) *Theoretical Soil Mechanics*. John Wiley and Sons, New York

Chapter 8
Tunnel Face Instability: Floresta Tunnels, Spain

TABLE OF CONTENTS

A.M. Puzrin et al., *Geomechanics of Failures*, DOI 10.1007/978-90-481-3531-8_8,

Chapter 8

Tunnel Face Instability: Floresta Tunnels, Spain

8.1 Field Observations

8.1.1 Tunnel face instability

In 1989 two short, parallel highway tunnels, 250 m long (East and West tunnels), were under construction north of Barcelona (La Floresta), under moderate covers (less than 40 m). The tunnels were driven through highly tectonized Paleozoic shales. Figure 8.1 shows a plan view of the two tunnels located at a distance of 35 m between axes. The upper half section of the tunnel, having a maximum height of 6 m, was mechanically excavated by means of boom headers.

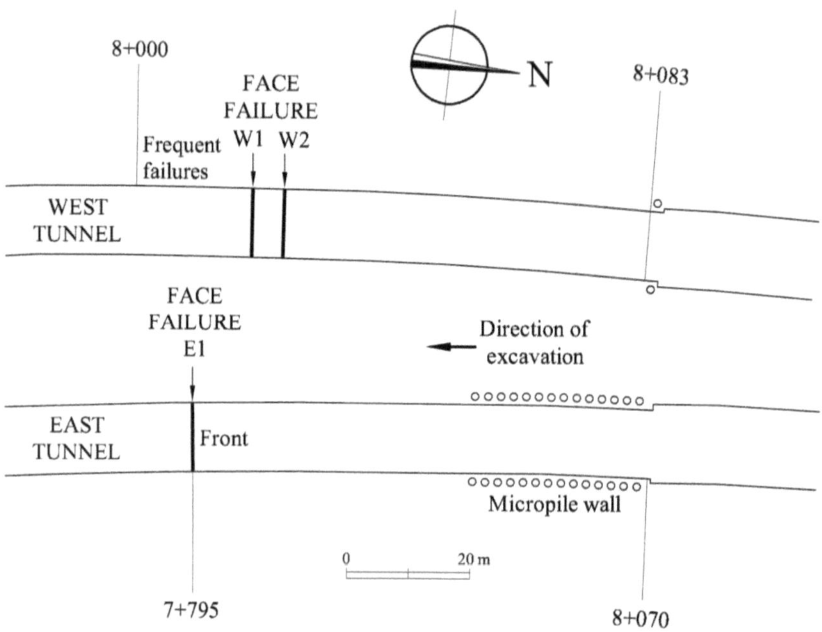

Figure 8.1 Plan view of the Floresta tunnels.

The tunnels were supported by heavy HEB 180 steel sets, closely spaced (1 m), as well as shotcrete and wire mesh (Fig. 8.2a). In addition, a continuous mass concrete lining, cast against perforated "Bernold" steel plates, supported by the HEB beams, closely followed the excavation face (Fig. 8.2). Excavation advanced from north to south. Face failures occurred almost simultaneously in both tunnels when the excavated distance from the portals was 60 m in the East tunnel and 100 m in the West tube. Figure 8.2b shows the West tunnel face failure. Ground cover was 22 m when the face of the East tunnel collapsed and 25.5 – 27 m for the two successive failures of the West tunnels. In what follows, these face failures will be named E1 (East tunnel) and W1, W2 (West tunnel).

(a)

(b)

Figure 8.2 (a) View of an unstable face in West tunnel; (b) tunnel face failure.

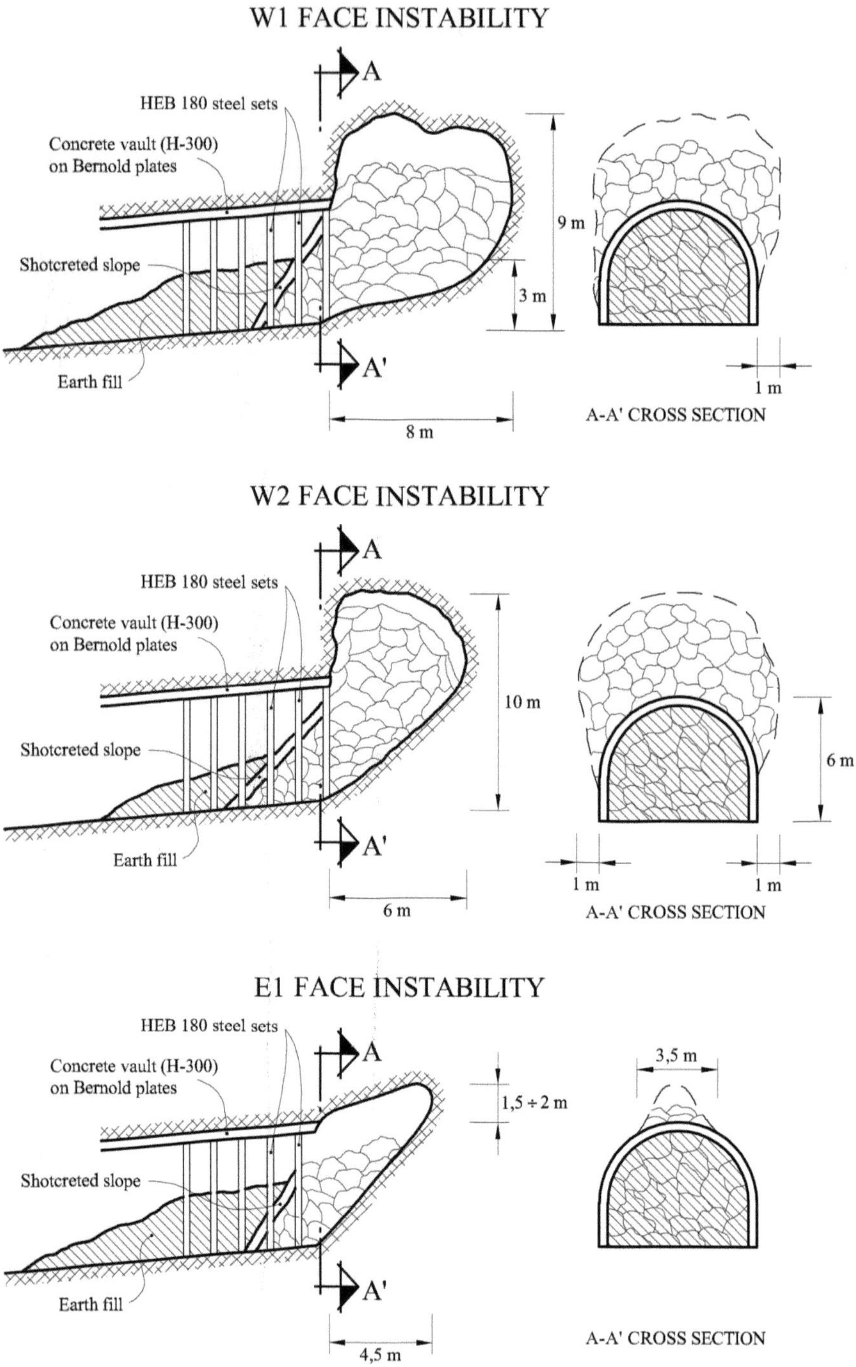

Figure 8.3 Sketches of the tunnel face failures reported by the engineer in charge.

Tunnel design documents characterized the Paleozoic shales by means of RMR indices (Bieniawski, 1989). Three main joint families were identified. In addition, several fracture directions and faults were also reported. Finally, the planes of schistosity contributed to a densely fissured mass. In fact, when all the reported discontinuities were represented in a common stereo net, a uniform distribution of discontinuities, having every possible dip, emerged. RMR values as low as 15 – 20 were reported for the faulted zones. Higher values (30 – 40) were assigned to shale with some quartzitic interbedding. However, a direct examination of the shale after excavation led to a more pessimistic rating with maximum values in the range 7 – 33 and minimum values close to zero!

Visual observations of tunnel faces after the failures revealed an extremely fractured and folded shale mass. The material could be easily broken by hand and exposed schistosity surfaces felt greasy to the touch, similar to talc surfaces.

The collapsed faces of the tunnels could be inspected in some detail. Sketches prepared by the engineer in charge are reproduced in Figure 8.3. Schistosity planes could no longer be recognized in the failed mass, which could be described as a soil. Discontinuity planes on the exposed failed face, shined when illuminated. No free water, however, was observed.

After the failures, the invading shale mass remained stable and adopted a small slope which is represented in Figure 8.3. The exposed surface of the failed mass was immediately shotcreted and additional fill was also accumulated against the tunnel face to improve stability conditions.

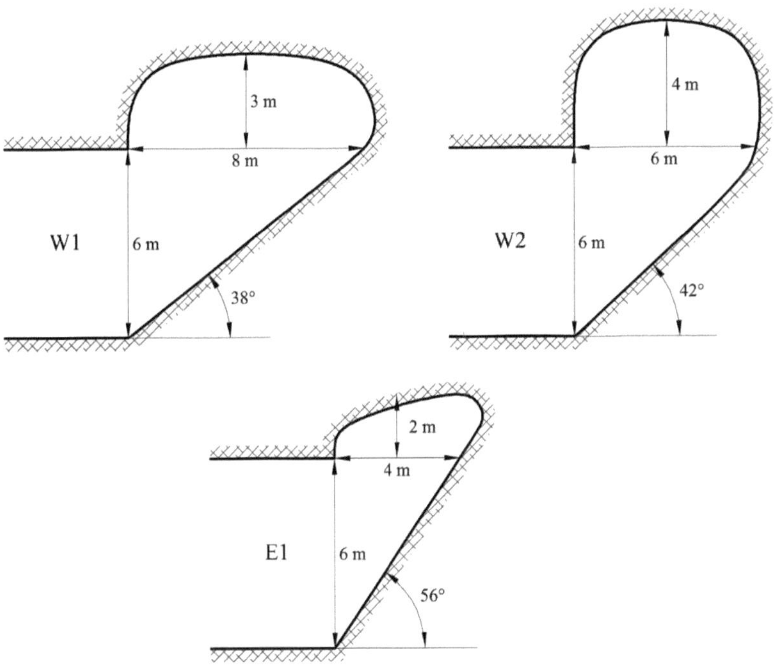

Figure 8.4 Longitudinal cross-sections of face failures.

(a)

(b)

Figure 8.5 Portal excavation: (a) the exposed highly fractured shale is immediately shotcreted after excavation; (b) slope failure.

Based on these field sketches, the geometry of the collapsed face and the size of the generated caverns are represented in cross-section in Figure 8.4. The two failures of the West tunnel (W1 and W2) affected the full face of the tunnel. In the East tunnel, the face collapse was more limited.

In this chapter the observed face failures are first examined and explained with the help of classical solutions for tunnel face stability under undrained conditions. In order to proceed with the construction of the tunnel, it was proposed to protect the excavation with micropile umbrellas built from the tunnel face. The improved stability conditions associated with this solution are developed in the chapter. Finally, the design of the umbrella actually installed is made on the basis of the analysis performed.

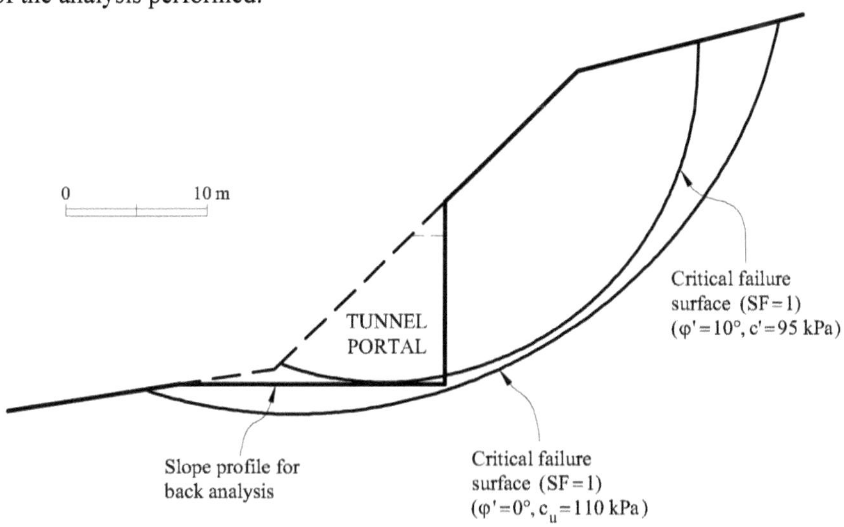

Figure 8.6 Slope instability in portal of the East tunnel.

8.1.2 Portal instability

The face failures were not the sole incident in the tunnelling work. In fact, a slope stability failure was triggered during the excavation of the portal of the East tunnel. A view of the collapsed slope (shotcreted in part) is shown in Figure 8.5. Figure 8.6 shows a cross-section of the slope and the vertical excavation performed to initiate tunnelling operations. An approximate circular failure mechanism could be inferred from the geometry after failure. This instability offered a good opportunity to estimate the mass strength parameters of the tectonized shale. A back-analysis of an "undrained" failure lead to a critical strength c_u = 110 kPa (the critical circular failure surface for this case is plotted in Fig. 8.6). If the friction angle is taken as $\varphi' = 10°$, the effective cohesion, which explains the failure, is $c' = 95$ kPa. The critical circle in this case is also plotted in Figure 8.6 (it is somewhat more "shallow"). Both are essentially consistent with field observations, although eye witnesses of the failure insisted on the deep nature of the observed failure. Water pressure was not considered in the second drained analysis, because no indication of water in the field was reported.

It appears, therefore, that this highly fractured and low friction shale could be approximated by a cohesive homogeneous material, when interpreting mass failures. Given the advantages of performing undrained analysis of the stability of tunnel faces, this hypothesis will be maintained in the rest of the chapter.

8.2 Tunnel Face Failures Explained by Plasticity Solutions

Consider first the classic work of Davis *et al.* (1980) on the stability of shallow tunnels in cohesive materials. Face stability was approached by means of the bound theorems of plasticity which are described in detail in Atkinson (1981) and Chen and Liu (1990).

The tunnel geometry and loading conditions analyzed by Davis *et al.* (1980) are shown in Figure 8.7a. *C, D* describe the cover and tunnel diameter, *P* is the non-supported length close to the tunnel face (*P* has been taken as zero in our case) and σ_S and σ_T are the stresses applied on the tunnel surface and against the face, respectively. $\sigma_S = \sigma_T = 0$ represents the actual loading conditions of the excavation of the two Floresta tunnels (no tunnel face support, no surface stresses applied).

Solutions given by Davis *et al.* (1980) were given in terms of a dimensionless "stability number", *N*, defined as follows:

$$N = \frac{\sigma_S - \sigma_T + \gamma(C + D/2)}{c_u}, \tag{8.1}$$

where γ is the natural specific weight of the soil. The numerator may be identified as the stress which tends to make the tunnel face unstable. The denominator is the available resisting strength.

Davis *et al.* (1980) provided upper and lower bounds for the *N* values in terms of *C/D*. If *N* is known, Equation (8.1) may be used to determine the c_u value which explains the face failure.

For instance, if a two-dimensional plane strain analysis is accepted, Table 8.1 summarizes the c_u values calculated for the three face failures described above.

Table 8.1 c_u values derived from the three face failures of the Floresta tunnels.

Face failure	Cover C (m)	Tunnel diameter D (m)	Upper bound theorem		Lower bound theorem			
					Smooth lining		Rough lining	
			$N^{(1)}$	c_u (kPa)	$N^{(1)}$	c_u (kPa)	$N^{(1)}$	c_u (kPa)
E1	22	6	7.9	72.6	5.2	115	6	96
W1	27	6	8.7	79	5.5	128	6.2	111
W2	25.5	6	8.5	77	5.4	119	6.1	107

A detailed derivation of the stability conditions associated with the upper bound theorem when reinforcing steel micropiles are introduced, is given later. The upper bound solution described by Davis *et al.* (1980) corresponds to the case of no reinforcing. Both cases are analyzed in the Appendix 8.2.

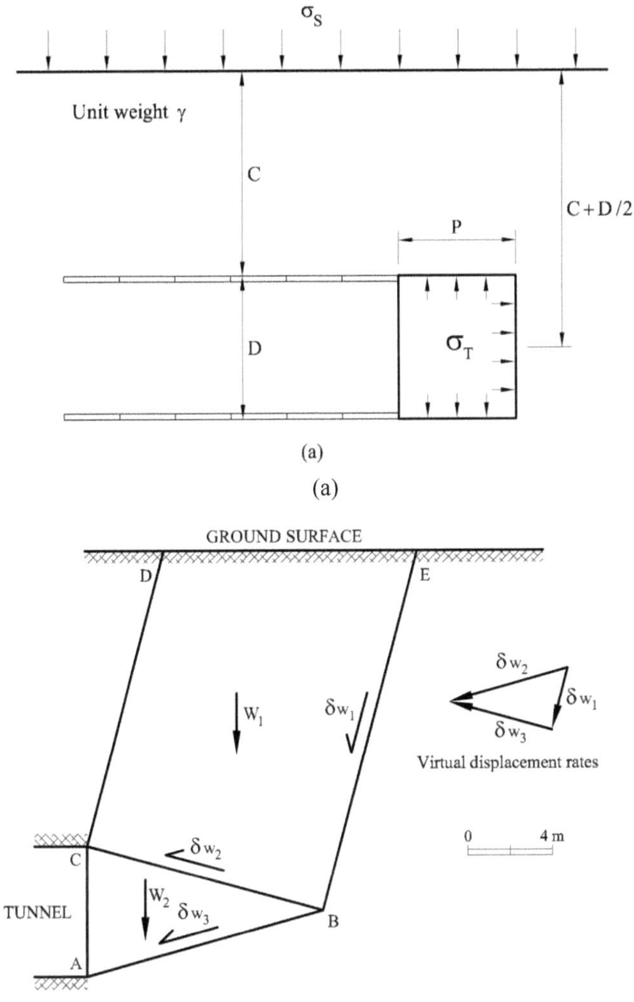

Figure 8.7 (a) Geometry and loading conditions of shallow tunnels considered in Davis *et al.* (1980); (b) Collapse mechanism in the Floresta tunnels. The set of virtual displacement rates are represented at point B.

The exact value of N is bracketed between the high values provided by the upper bound theorem and the low values provided by the lower bound theorem (see Table 8.1). The optimization process leads to the reduction of the $N_{\text{upper bound}}$ and an increase of the $N_{\text{lower bound}}$. Since c_u is inversely related to N (Eq. (8.1)) the c_u value associated with the upper bound decreases during the optimization process. Therefore c_u obtained through the upper bound provides a lower bound for the undrained strength. For analogous reason, the lower bound theorem provides an upper bound of c_u.

Calculations, summarized in Table 8.1, were made for $\gamma = 23$ kN/m^3 and the solutions given by Davis *et al.* (1980) were used to find the N parameters. Then c_u is derived from Equation (8.1) as

$$c_u = \frac{\gamma(C + D/2)}{N}. \tag{8.2}$$

It is satisfactory that the c_u values derived from the stability analysis of the face failures are very close to the value derived from the back-analysis of the portal failure described above. This consistency adds some confidence to the overall analysis performed.

The application of the upper bound theorem requires the specification of a family of failure geometries. The minimization process involved in the application of the theorem provides the critical mechanism (for the adopted family of failure mechanisms). The critical failure mechanism, for the East tunnel failure, is plotted in Figure 8.7b. The mechanism is built by means of a triangular wedge, which may slide into the tunnel face, and a second rectangular wedge loading the upper face of the triangular wedge. The soil strength, c_u, develops whenever there is a differential motion between any two surfaces in contact. Therefore, in the case represented in Figure 8.7b, c_u acts along surfaces AB, BC, BE and CD. Also indicated in the figure are the rigid body virtual displacement rates (also called "velocities") of the two moving "wedges" (weights W_1 and W_2): δ_{W_1}, δ_{W_3} and δ_{W_2}.

Once the first failures occurred further excavation implied an increasing cover (C) and therefore an increasing risk of face failure because higher values of c_u for equilibrium are required (Eq. (8.2)). If the c_u value derived from the back-analysis of the portal failure represented "in situ" conditions, it is clear that face instability could not be avoided.

Face stabilization may be achieved by a few alternative procedures:

- reducing the size of the unsupported face by means of a soil buttress (Fig. 8.8a);
- reinforcing the soil ahead of the face by means of injected bolts (they are typically made of glass fibre in order to later facilitate its excavation) (Fig. 8.8b), or
- installing an umbrella of closely spaced micropiles, drilled from the tunnel (Fig. 8.8c).

The third solution, which was selected by the contractor, will be further developed here.

8.3 Face Stabilization by Subhorizontal Micropiles

A first step in the evaluation of this solution is the analysis of the behaviour of a subhorizontal micropile subjected to the expected failure mechanism. Plane strain and undrained conditions will be assumed.

Consider the previous failure mechanism (Fig. 8.7), when a micropile crosses the upper wedge. The micropile will react against the expected displacement imposed by the mechanism (Fig. 8.9a). At point P, the micropile action on the wedge will be characterized by a normal force, N, a shear force, Q, and a moment, M. If the wedge failure mechanism is maintained (a sliding displacement of the wedge), moment M will not produce any external (stabilizing) work. Only Q and N will contribute to increasing safety. However, M must be determined because the actual values of N and Q will be controlled by the failure criterion of the steel section of the micropile. In fact, failure criteria are given in terms of stresses (normal and shear), which depend, not only on shear and normal forces (Q, N), but also on the bending moment, M.

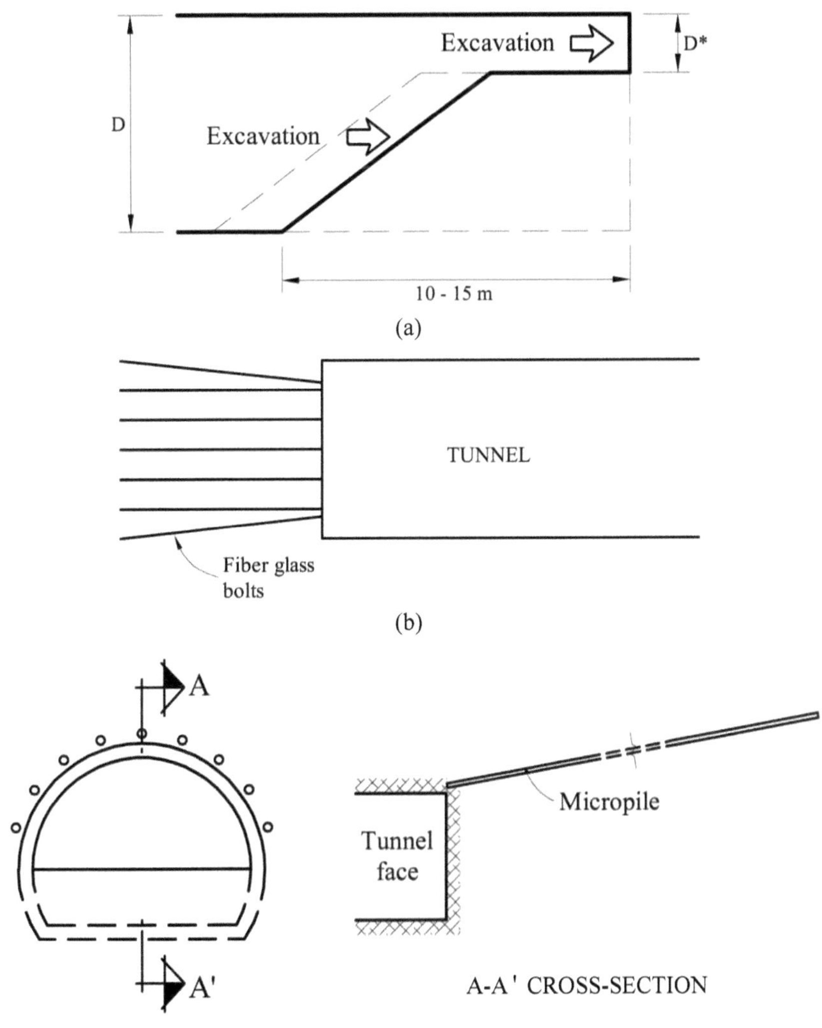

Figure 8.8 (a) Soil buttress against the face; (b) face reinforcement; (c) micropile umbrella to prevent face instability.

In order to include the micropile effect into a stability analysis of the tunnel face, the subsequent approach will be followed:

- A failure mechanism will be analyzed. The analysis will follow an upper bound plasticity approach. The micropile effect will be substituted by the limiting normal (N) and shear (Q) forces at point P, in which the micropile crosses the sliding surface of the most critical mechanism.
- Forces exerted by the micropile on the critical wedge will be determined in an independent analysis of the micropile, considered as a beam and subjected to the kinematic motion imposed by the assumed failure mechanism. The micropile critical section (point P) will be taken to critical conditions. In other words, the steel section will plastify.
- The most critical mechanism will be determined by means of an optimization procedure, based on the tenets of the upper bound theorem of perfect plasticity. Micropile forces (Q and N) will be considered as external forces on the mechanism.

8.3.1 Beam behaviour. Limiting conditions

Consider, in Figure 8.9b, a micropile isolated from the surrounding soil. The length of the beam included in the moving wedge will experience a uniform displacement, δ, in the direction indicated. This displacement has two components: δ_v, normal to the beam, and δ_h, in the direction of the beam axis. δ_v will induce shear forces and bending moments on the beam. δ_h will result only in normal forces. In order to simplify the calculation of the beam, the embedded length of the micropile into the stationary soil will be assumed to be equivalent to a beam of length, b, between the crossing point P and a fixed (fully clamped) point X. A procedure to approximate the equivalent beam length, b, will be given later.

For a perfectly plastic material (steel in most cases) obeying Von Mises' yield criterion, the relationship satisfied by the normal (σ) and shear stress (τ) acting on any plastified fiber in a cross-section is (Appendix 8.1):

$$\sigma^2 + 3\tau^2 = \sigma_e^2, \tag{8.3}$$

where σ_e is the uniaxial tensile strength or elastic limit of the micropile material. This condition will be imposed on section P (see Fig. 8.9a); σ and τ will be expressed in terms of the normal (N) and shear (Q) forces, as well as the bending moment (M).

Once these forces are calculated (in terms of the imposed displacement, δ), the stresses (normal and shear) in each fiber of the section can be obtained. Substituting the stress values of the first fiber to reach yielding (that is, the fiber in which the value of $\sigma^2 + 3\tau^2$ is a maximum) into Equation (8.3) will allow determination of the critical displacement, δ. It will be accepted that this displacement leads to the maximum support provided by the micropile (see, however, the discussion at the end of the chapter). Then, forces N and Q, resisting the failure mechanism, could be found. The final step will be the application of the upper limit theorem of plasticity in order to derive a collapse load. Micropile resisting loads N and Q will now play the role of external loads.

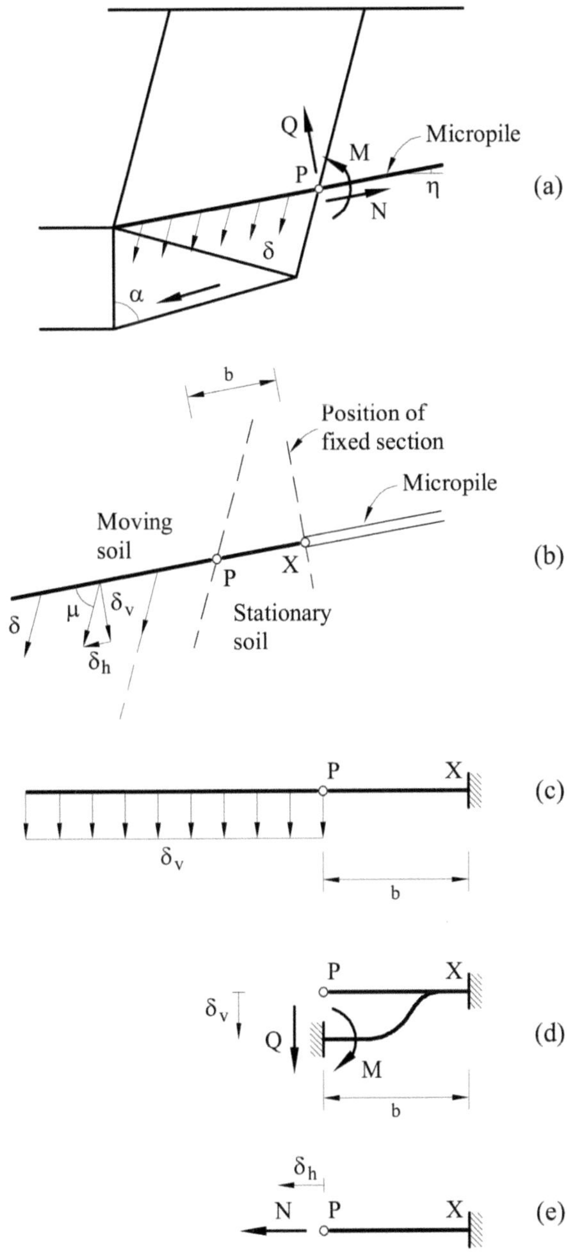

Figure 8.9 (a) Micropile actions on failure mechanism; (b) uniform displacement δ on a finite length of the micropile; (c) and (d) bending behaviour of the micropile; (e) tensile behaviour of the micropile.

The eventual support of the micropile on the tunnel side will be disregarded and, therefore, micropiles will be assumed to work as cantilever beams, fixed at some point in the stable ground ahead of the stable mechanism. Consider now the equivalent beam fixed on one end, represented in Figures 8.9c,d,e. The bending moment and shear force on point P, induced by a relative displacement, δ_v, of one extreme with respect to the other are given by

$$M=\frac{6EI_x}{b^2}\delta_v, \tag{8.4}$$

$$Q=\frac{12EI_x}{b^3}\delta_v. \tag{8.5}$$

These expressions may be found in manuals of strength of materials (see, for instance, Young, 1989). The normal force (Fig. 8.9e) is also easily found as

$$N=\frac{AE}{b}\delta_h. \tag{8.6}$$

In the preceding equations, E is the steel elastic modulus, while I_x and A are the moment of inertia with respect to the horizontal axis of the section and the cross-sectional area of the micropile, respectively. δ_v and δ_h are now written in terms of the angle μ between the direction of the imposed displacement and the direction normal to the micropile (Fig. 8.9b):

$$\delta_v=\delta\sin\mu \ ; \ \delta_h=\delta\cos\mu. \tag{8.7}$$

Normal stresses due to the force pair (σ_N, σ_M) are given by (Fig. 8.10a):

$$\sigma=\sigma_N+\sigma_M=\frac{N}{A}-\frac{Mz}{I_x}, \tag{8.8}$$

where z is the distance from the beam axis, x, to a particular point of the section. Negative values indicate compression.

If expressions for the normal force (Eq. (8.6)) and bending moment (Eq. (8.4)) are substituted into Equation (8.8):

$$\sigma=\left(\frac{E}{b}\cos\mu-\frac{6Ez}{b^2}\sin\mu\right)\delta. \tag{8.9}$$

Shear stresses (Fig 8.10b,c), due to shear force (Q), are given at any point of the cross-section characterized by a coordinate z, with respect to the mean plane by

$$\tau=\frac{2Q\sqrt{d^2-4z^2}}{\pi d^2 t}, \tag{8.10}$$

where t is the thickness of the steel tube and d the diameter of the micropile.

Substituting Q, given by Equation (8.5), Equation (8.10) results in

$$\tau = \frac{3Ed}{b^3}\delta \sin\mu\sqrt{d^2 - 4z^2}. \tag{8.11}$$

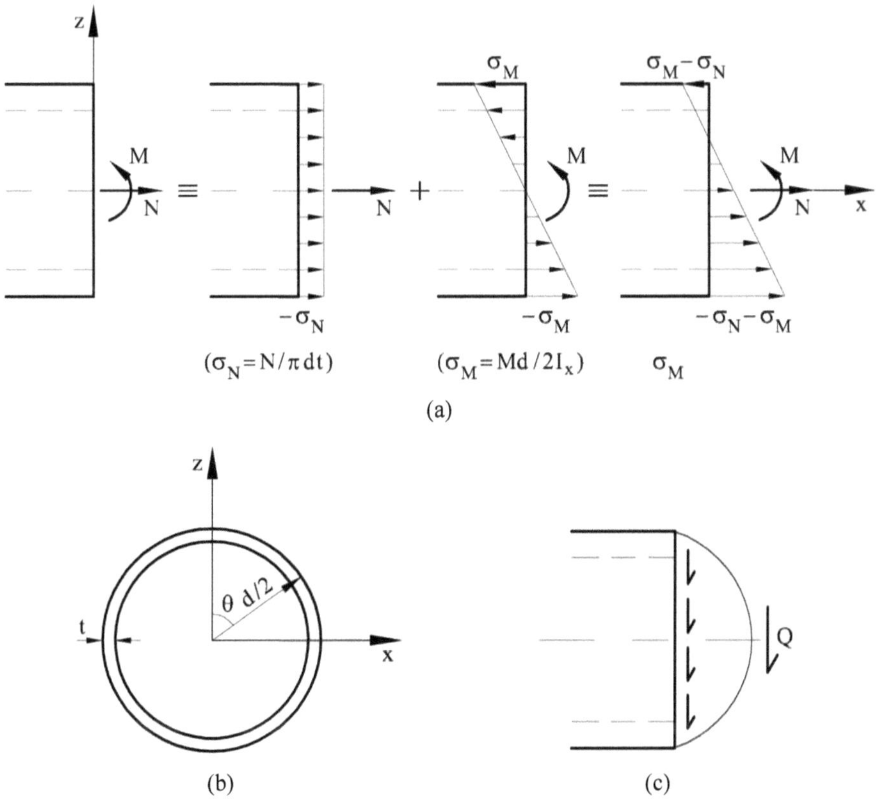

Figure 8.10 Stress distributions (a) σ induced by normal force (N) and bending moment (M) and (c) τ induced by shear force (Q) in the (b) micropile cross-section.

A conservative assumption will be introduced now in the calculation of the available strength provided by the micropile. The strength will be calculated as the value associated with the state in which the section starts to yield at some point (or "fiber"). Steel yielding may continue beyond this point; however this increment of strength in not considered here. The first section fiber to yield is given by the coordinate z which provides the maximum value of the Von Mises' stress:

$$\sigma^2 + 3\tau^2 = \left(\frac{E}{b}\cos\mu - \frac{6Ez}{b^2}\sin\mu\right)^2 \delta^2 + 27\left(\frac{Ed}{b^3}\delta\sin\mu\sqrt{d^2 - 4z^2}\right)^2, \tag{8.12}$$

where the equations for the normal and shear stresses (Eqs. (8.9) and (8.11)) have been introduced.

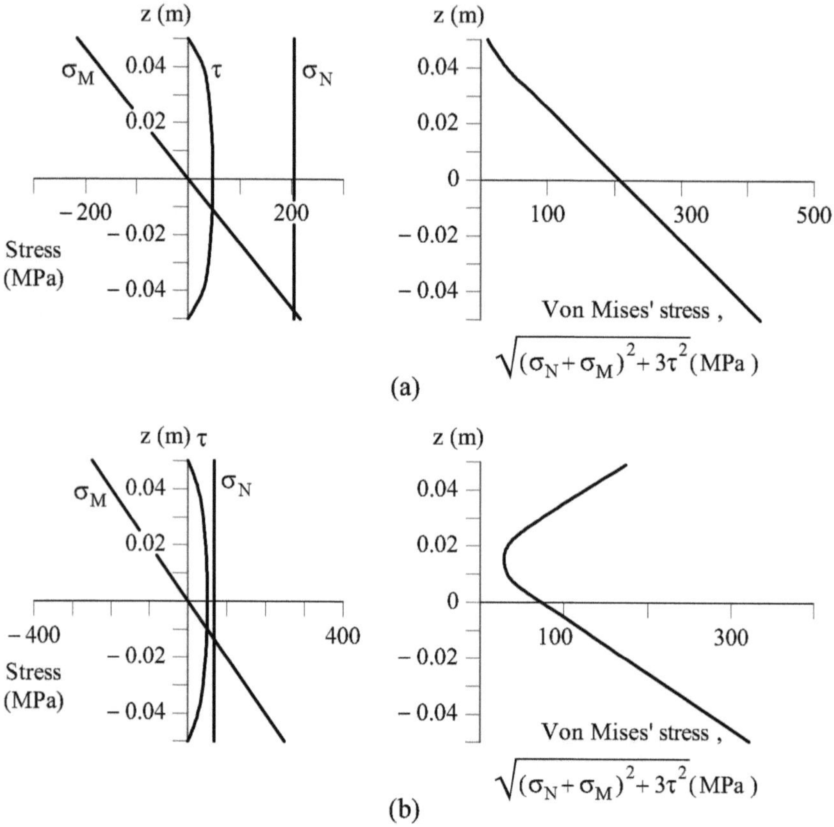

Figure 8.11 Calculated normal and shear stresses on the micropile cross-section and the corresponding Von Mises' stress for (a) α = 50° and (b) α = 70° (angle α is defined in Fig. 8.9a).

Figure 8.11 shows the distribution of σ_N, σ_M and τ along the cross-section at point P (Fig. 8.9) for some characteristic micropile parameters indicated between parentheses in Table 8.2 and for a unit displacement. Micropile inclination with respect to the horizontal has been assumed equal to 10° and the steel Young modulus is 210 GPa. Two angles α, which defines the collapse geometry (see Fig. A8.3) have been considered (50° and 70°).

Note that the values of τ reach a maximum in the center of the section. However, the sum of σ_N and σ_M in the most critical position ($z=-R$) provides a high σ tensile stress if compared with τ. It turns out that $\sigma^2+3\tau^2$ reaches a maximum at $z=-R$. Bending dominates the tensile stressing of the micropile.

However, this conclusion is tied to the particular problem we are considering. It is associated with the particular cross-section of the micropile (a hollow steel cylinder having a relatively thin thickness) and the imposed loading mechanism.

Table 8.2 Range of parameters for micropile reinforcement. Values for the umbrella designed for La Floresta tunnels are also indicated in brackets.

Parameter	Symbol	Unit	Range of values (La Floresta)
Beam diameter	d	m	0.04 – 0.12 (0.1)
Beam thickness	t	m	0.003 – 0.011 (0.01)
Distance between micropiles	s	m	0.2 – 1 (0.2)
Steel strength	σ_e	MPa	200 – 500 (400)
Soil undrained strength	c_u	MPa	0.03 – 0.5 (0.07)
Tunnel diameter	D	m	2 – 12 (6)

If shear and normal stresses (Eqs. (8.9) and (8.11)) at the fiber $z=-R$ are calculated and the Von Mises' criterion (Eq. (8.3)) is taken into account the following expression for the displacement δ leading to the first yield of a fiber in the micropile cross section is derived:

$$\delta=\frac{\sigma_e b}{E}\frac{1}{\sqrt{f(d/b,\mu)}}, \tag{8.13}$$

where $f(d/b,\mu)$ is a function of the ratio d/b and the relative orientation of the micropile and the upper sliding wedge ($\mu=\alpha-\eta$; see Fig. 8.9a), given by

$$f(d/b,\mu)=6\cos\mu\sin\mu(d/b)+9\sin^2\mu(d/b)^2+\cos^2\mu. \tag{8.14}$$

When the δ value given in Equation (8.13) is substituted in Equations (8.5) and (8.6) (considering Eq. (8.7)), the following shear and tensile forces applied by the micropile on the sliding mechanism, at point P are found:

$$N=\sigma_e td\pi\frac{\cos\mu}{\sqrt{f(d/b,\mu)}}, \tag{8.15}$$

$$Q=\frac{3}{2}\sigma_e tb\pi(d/b)^3\frac{\sin\mu}{\sqrt{f(d/b,\mu)}}. \tag{8.16}$$

8.3.2 Analysis of the beam-reinforced collapse mechanism

The upper bound theorem of plasticity was applied in the manner detailed in the Appendix 8.2. The standard solution for the two wedges, a plain strain mechanism under undrained conditions, was first derived as an introduction to the reinforced case. The resisting normal and shear forces exerted by the micropiles (Eqs. (8.15) and (8.16)) on the inner boundary of the collapse mechanism were introduced as external forces. The minimization process implied in the procedure is now more involved because N and Q depend on the geometry of the mechanism.

The application of the upper bound theorem of plasticity requires the calculation of the plastic external work, which is then made equal to the internal dissipation work. This procedure, detailed in Appendix 8.2, leads to the following equation:

$$\frac{(\sigma_S-\sigma_T)}{c_u}-\frac{1}{\tan\alpha}\left(\frac{4C}{D}+1\right)-\tan\alpha+\frac{\gamma D}{c_u}\left(\frac{C}{D}+\frac{1}{2}\right)-\frac{\sigma_e td}{c_u sD}\frac{2\pi\cos\alpha(\cos^2\mu+\frac{3}{2}\left(\frac{d}{b}\right)^2\sin^2\mu)}{\sqrt{f\left(\frac{d}{b},\mu\right)}}=0. \tag{8.17}$$

Any set of parameters that satisfies the previous equation describes a collapse mechanism. The first three terms identify the upper bound expression in the absence of reinforcement. The reinforcement is identified by the dimensionless parameter $\sigma_e td/c_u sD$ which combines, in a single expression, the tensile strength of the steel (σ_e), the diameter (d) and thickness (t) of the tubular reinforcement, the soil undrained strength (c_u), the distance between axis of micropiles (s) and the tunnel diameter (D). We will refer to this coefficient as the "Micropile Coefficient".

If a solution is sought for the normalized stress difference, $(\sigma_S-\sigma_T)/c_u$, the following form of the upper bound theorem is found:

$$\frac{(\sigma_S-\sigma_T)}{c_u}\le\frac{1}{\tan\alpha}\left(\frac{4C}{D}+1\right)+\tan\alpha-\frac{\gamma D}{c_u}\left(\frac{C}{D}+\frac{1}{2}\right)+\frac{\sigma_e td}{c_u sD}\frac{\pi\cos\alpha(2\cos^2\mu+3\left(\frac{d}{b}\right)^2\sin^2\mu)}{\sqrt{f\left(\frac{d}{b},\mu\right)}}. \tag{8.18}$$

In this case, the term on the right will be minimized with respect to the angle α in order to obtain the critical mechanism that leads to the smallest of the upper bound solutions. Note that μ is a function of α. The development of this solution can also be found in the Appendix 8.2.

In the absence of reinforcement, the upper bound solution given by Davis *et al.* (1980) may be represented in the manner shown in Figure 8.12. In this graph, the minimized value of $(\sigma_S-\sigma_T)/c_u$ with respect to angle α is plotted against the cover ratio C/D, for different values of the strength ratio $\gamma D/c_u$. This representation is preferred here over the stability ratio N, selected by Davis *et al.* (1980), because it is more convenient when dealing with the reinforced case developed later. (The same representation is followed by Augarde *et al.* (2003) who provided solutions for non-homogeneous strength profiles.) The critical angle α obtained for the unreinforced case is plotted in Figure 8.15.

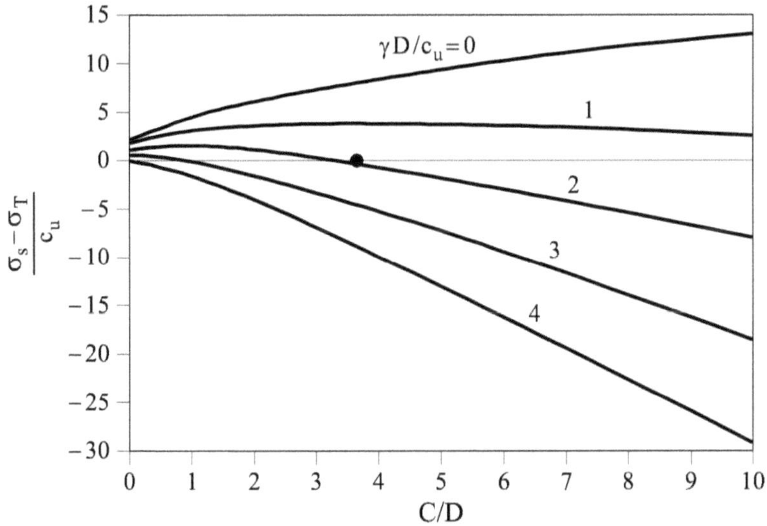

Figure 8.12 Upper bound solutions for the unreinforced tunnel face. Plane strain, undrained (based on the results presented by Davis *et al.*, 1980). The black dot represents conditions of the East Floresta Tunnel.

The solution for the reinforced tunnel face can be obtained in a similar way. In this case, the last term of Equation (8.20) is included in the derivation (see Appendix 8.2).

In Table 8.2, typical parameters of micropiles used in practice for tunnel reinforcement are given. Other parameters defining the dimensionless parameter $\sigma_e td/c_u sD$ have also been introduced. They reflect reasonable bounds of current practice. The calculated values of parameter $\sigma_e td/c_u sD$ for extreme cases may range between 0.1 (weak reinforcement) and 500 (strong reinforcement). Figures 8.13 and 8.14 show the upper bound solution of $(\sigma_S - \sigma_T)/c_u$ for two values of $\sigma_e td/c_u sD$: 5 and 40. The adopted distance, *b*, for the position of the "clamped" section of the micropile (Fig. 8.9b) is five times the micropile diameter (more will be said later about this choice). Additional assumptions are the micropile inclination with respect to the horizontal ($\eta = 10°$) and the steel Young modulus (210 GPa). The upper bound values of $(\sigma_S - \sigma_T)/c_u$ depend on $\gamma D/c_u$ and C/D, as in the unreinforced case. To visualize better the effect of reinforcement, the unreinforced case is also plotted in Figures 8.13 and 8.14. Critical angles associated with the plotted cases are shown in Figure 8.15.

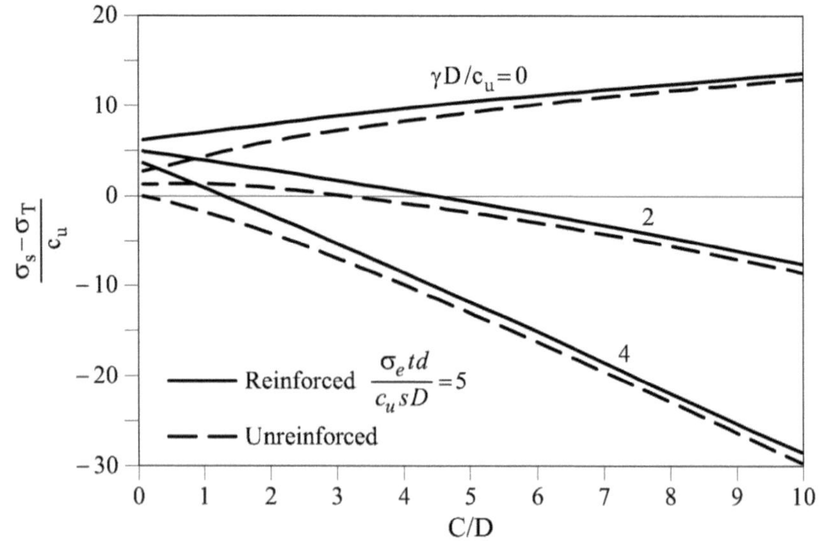

Figure 8.13 Upper bound solutions for the reinforced tunnel face ($\sigma_e td/c_u sD = 5$) and the unreinforced case. Plane strain, undrained ($d/b = 0.20$).

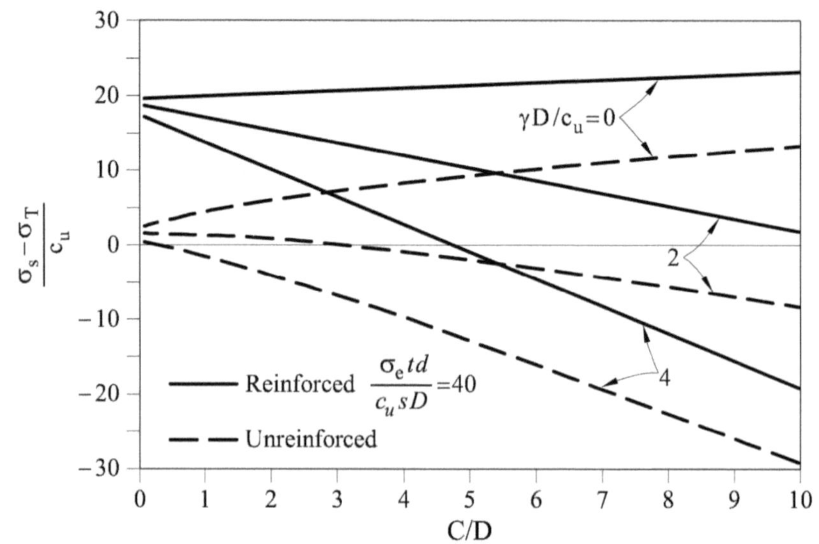

Figure 8.14 Upper bound solutions for the reinforced tunnel face ($\sigma_e td/c_u sD = 40$) and the unreinforced case. Plane strain, undrained ($d/b = 0.20$).

Note that the reinforcement leads to wider wedges (higher α values), if compared with the standard solution.

The solution developed so far minimizes the external “load”, $(\sigma_S - \sigma_T)/c_u$, against the geometry of the failure mechanism, accepting that the micropile action is known (Eq. (8.16)). But the normal and shear forces exerted by the micropile on the failure mechanism are also external loads and the upper bound theorem may be also applied to the critical micropile stabilizing action, assuming that the remaining external loads are known. In fact, the Micropile Coefficient can be isolated from (8.17):

$$\frac{\sigma_e td}{c_u sD} \leq \frac{\sqrt{f\left(\frac{d}{b},\mu\right)}\left[\frac{(\sigma_S-\sigma_T)}{c_u}-\frac{1}{\tan\alpha}\left(\frac{4C}{D}+1\right)-\tan\alpha+\frac{\gamma D}{c_u}\left(\frac{C}{D}+\frac{1}{2}\right)\right]}{2\pi\cos\alpha(\cos^2\mu+\frac{3}{2}\left(\frac{d}{b}\right)^2\sin^2\mu)}. \tag{8.19}$$

A maximum value of $\sigma_e td/c_u sD$, with respect to α, is found in this case in order to obtain the collapse mechanism exhibiting the maximum support provided by the micropile. Any micropile having lower mechanical properties will result in a collapse.

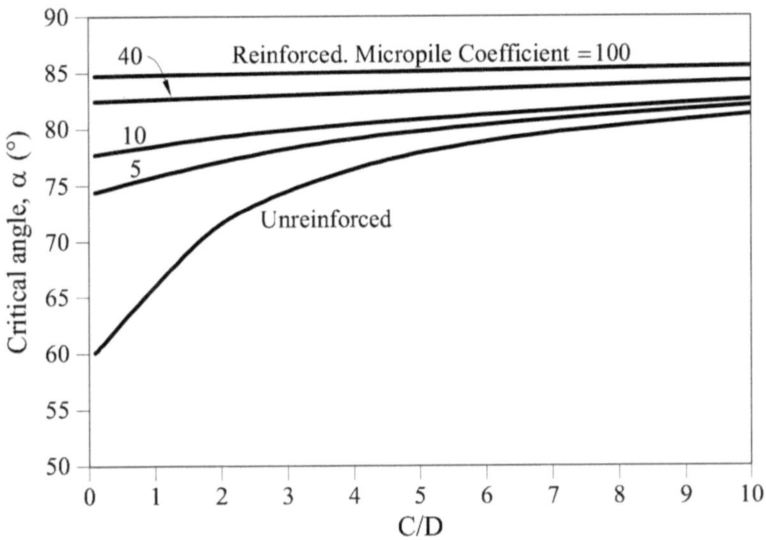

Figure 8.15 Values of critical angle α for unreinforced and reinforced case ($d/b = 0.20$).

The critical values of the Micropile Coefficient have been found and plotted in Figures 8.16 and 8.17 (Micropile Coefficient in log scale) in terms of *C/D* and $\gamma D/c_u$, for the special case $(\sigma_S - \sigma_T)/c_u = 0$. This case is of particular importance for open face tunnelling (this is also the case of La Floresta tunnels). On the other hand, micropile stabilization of tunnel faces is, in fact, usually performed in open face tunnelling. A value $d/b = 0.20$ was considered in this case.

The plot in Figures 8.16 and 8.17 is particularly interesting for designing a micropile umbrella because it directly provides the Micropile Coefficient. Note that negative values are calculated for a low value of $\gamma D/c_u$ (c_u is large compared with γD: either a strong soil or a small tunnel diameter or both). They simply indicate that the micropile umbrella is not needed for stabilization.

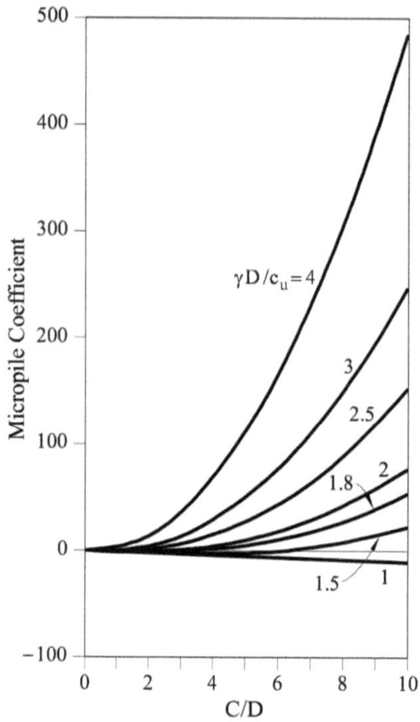

Figure 8.16 Upper bound solutions for the Micropile Coefficient. Plane strain, undrained. $(\sigma_S - \sigma_T)/c_u = 0$ and $d/b = 0.20$.

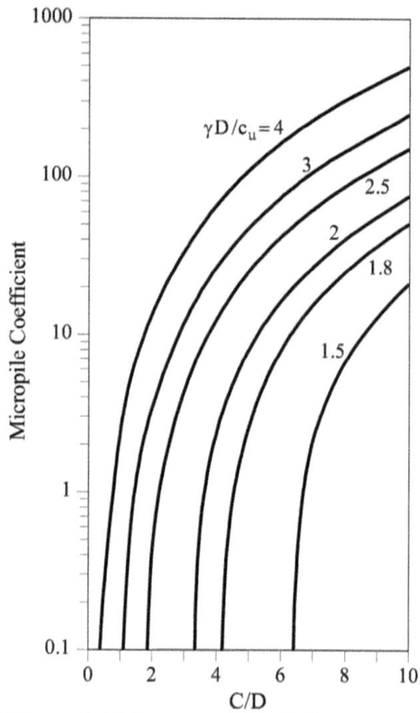

Figure 8.17 Upper bound solutions for the Micropile Coefficient. Plane strain, undrained. $(\sigma_S - \sigma_T)/c_u = 0$ and $d/b = 0.20$. Log scale for the Micropile Coefficient.

The two cases developed on the basis of the upper bound theorem, represented in Figures 8.13/8.14 and 8.16/8.17, are consistent. To show it, consider the following example: A tunnel face defined by $C/D = 8$ and $\gamma D/c_u = 2$, has to be stabilized by means of a micropile umbrella. Figures 8.16 or 8.17 (for $d/b = 0.20$) immediately provide the answer: The micropile design should satisfy $\sigma_e td/c_u sD = 40$. Consider now Figure 8.14, which corresponds to $\sigma_e td/c_u sD$=40: for a value $C/D = 8$ and $\gamma D/c_u = 2$ it provides $(\sigma_S - \sigma_T)/c_u = 0$, which is the hypothesis used in plotting Figures 8.16 and 8.17.

Figure 8.17 provides a more accurate representation of the solution, specially for low values of the Micropile Coefficient for the usual range of $\gamma D/c_u$ parameters found in practice.

8.3.3 Effect of relative clamping distance (*d*/*b*)

An estimation of the value of the clamping distance b (Fig. 8.9) may be obtained from the theory of piles embedded in an elastic half-space, subjected to a horizontal load and a moment at its head. This problem is described in Poulos and Davis (1980). Solutions depend on a "pile flexibility factor", K_R, defined as

$$K_R = \frac{E_p I_p}{E_s L^4}, \tag{8.20}$$

where E_p and E_s are the elastic moduli of pile and soil, I_p is the moment of inertia of the micropile and L its length. If some typical values are introduced (E_p = 210,000 MPa; I_p = 100 – 1,000 cm^4; E_s =10^2 – 10^3 MPa and L = 3 – 10 m), the resulting values of K_R are very small (10^{-6} to 10^{-8}). They correspond to a very "flexible" pile. In flexible piles subjected to horizontal load and moment, the pile displacements are confined to the vicinity of the head (a few solutions are given in Poulos and Davis, 1980). It can be accepted therefore that distance b will be relatively small: a few pile diameters.

Another classical solution for the same problem derives from the concept of modulus of subgrade reaction, k (Poulos and Davis, 1980). The "elastic length" of a beam of width d, inertia I_p and elastic modulus E_p is given by

$$L_{el} = \sqrt[4]{\frac{4E_p I_p}{kd}}. \tag{8.21}$$

Terzaghi (1955) proposed the following relationship to relate $k{\cdot}d$ with the undrained strength of soils:

$$kd = 66.7c_u. \tag{8.22}$$

If typical values for micropile reinforcement (Table 8.2) are introduced in Equation (8.21), the value of L_{el} for medium to hard cohesive soils may range between 75 and 125 cm. On the other hand, in piles subjected to horizontal load, the maximum moment develops at depths of the order of $L_{el}/2$ and $L_{el}/3$. Therefore, the ratio d/b will be typically in the range 0.2 to 0.1.

Additional solutions for the Micropile Coefficient for other values of the clamping distance have been found. They are plotted in Figure 8.18 for $d/b = 0.10$ and in Figure 8.19 for $d/b = 0.05$.

In order to appreciate better the effect of the relative clamping distance d/b, Figure 8.20 shows the variation of the micropile coefficient with *C*/*D* for different values of d/b and the particular case $\gamma D/c_u = 3$. The effect of d/b is seen to be small.

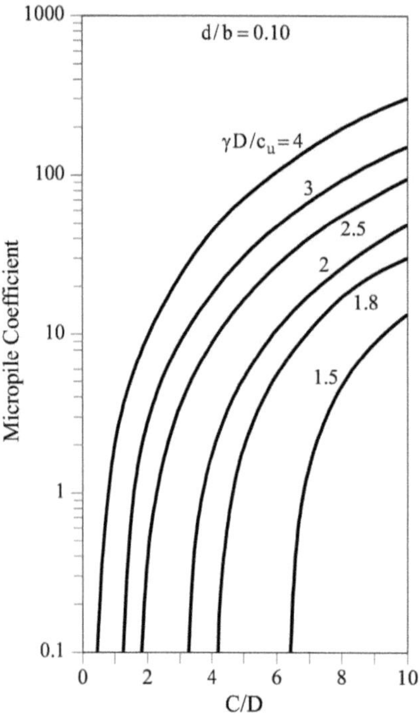

Figure 8.18 Upper bound solutions for the Micropile Coefficient. Plane strain, undrained. $(\sigma_S - \sigma_T)/c_u = 0$ and $d/b = 0.10$.

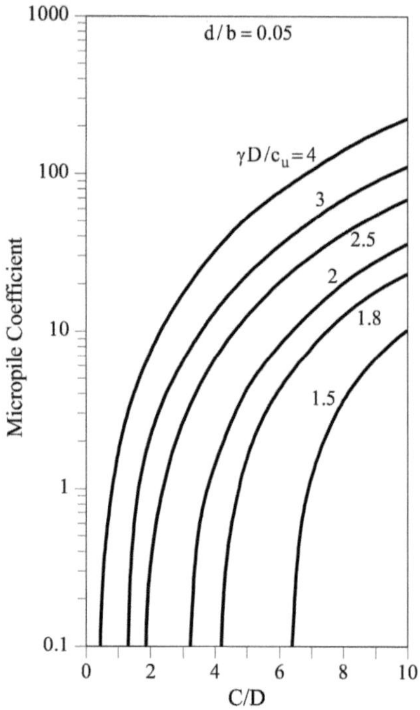

Figure 8.19 Upper bound solutions for the Micropile Coefficient. Plane strain, undrained. $(\sigma_S - \sigma_T)/c_u = 0$ and d/b = 0.05.

8.4 Reinforcing La Floresta Tunnels

Consider first the unstable situation. The East tunnel (C = 22 m; D = 6 m) is taken as a reference (the other two cases of instability in the West tunnel have similar dimensions). In Figure 8.12, in the absence of stresses applied in the surface and against the tunnel face, $(\sigma_S - \sigma_T)/c_u = 0$, and for $C/D = 3.66$ (indicated by a dot in Fig. 8.12), the following strength ratio is obtained: $\gamma D/c_u$ = 1.90. For an estimated soil unit weight of 23 kN/m^3, the calculated undrained strength which explains the tunnel face failure is $c_u = \gamma D/1.90$ = 72.6 kPa.

This value, which will be accepted as the field value of the undrained strength, has already been given in Table 8.1 on the basis of Davis *et al.* (1980) solutions.

Consider now the reinforcement selected for the face. It is desirable to calculate the achieved safety factor, *SF*, against failure of the face. The safety factor will be conventionally defined as the ratio between available shear strength (the field value) and the shear strength needed for equilibrium under design conditions.

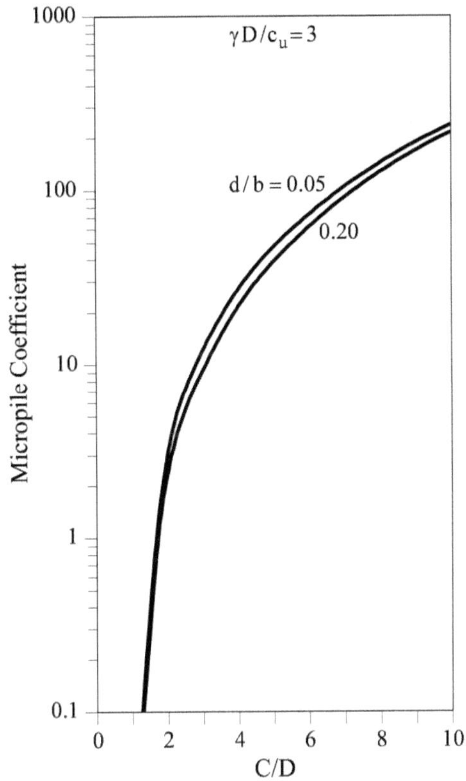

Figure 8.20 Effect of clamping distance on Micropile Coefficient. Plane strain, undrained, $(\sigma_S - \sigma_T)/c_u = 0$ and $\gamma D/c_u$ =3.

The latter is given by the application of the upper bound theorem. The calculation of the safety factor requires an iterative procedure because it cannot be isolated since both the Micropile Coefficient and the strength ratio depend on the c_u value required for equilibrium (in fact, the value provided by the application of the upper bound theorem). Consider, as a first trial, *SF* = 1.1. The corresponding strength to perform the calculations will be c_u = 72.6/1.1 = 66 kPa. Then, a Micropile Coefficient of 5 is calculated with the parameters indicated in Table 8.2. Figure 8.17, for d/b = 0.20 and *C/D* = 3.66, provides a strength ratio $\gamma D/c_u = 2.2$ and therefore an undrained strength c_u = $23 \times 6/2.2 = 62$ kPa, which is close to the value of 66 kPa. It is concluded that the proposed reinforcement maintains a safety factor in the vicinity of 1.1, which is perhaps too low.

Of course, the solution is not unique, in the sense that alternative geometrical characteristics of the micropile umbrella may result in the same value of the coefficient. Other considerations of a practical nature (availability of micropile tubes; boring diameter achieved with existing perforation rigs, reasonable spacing among micropiles, etc.) will dictate the final choice.

8.5 Discussion

The analysis presented relies in the application of the upper bound theorem of plasticity. Since the forces actually considered in the minimization process are resisting forces (the stress against the tunnel face or the resisting forces induced by the micropile) they oppose the collapse mechanism and its work is negative (unlike the external forces inducing collapse in a regular foundation case). This explains why the application of the “upper” bound theorem actually leads to resisting forces smaller than the actual ones. It gives an “unsafe” bound (just as in regular foundations where the upper bound theorem also produces an “unsafe” ultimate load since the true bearing capacity is lower than the predicted one!).

A further limitation comes from the two-dimensional, plane strain character of the problem solved. This second assumption lies on the conservative side, however (adding the third dimension would imply reduced stabilizing forces).

Another important source of conservativeness is the limit analysis of the micropile steel section. The assumption was that the limiting strength offered by the beam was given by the application of the Von Mises strength criterion to the “worst fiber” of the beam cross-section. Steel yielding may continue, however, beyond this point but the analysis required is outside the purpose of this chapter. The increase in limiting steel forces for a fully plastified case may amount to $15-17\%$ over the values calculated here. On the other hand, such refinement would not be justified in view of the set of limitations of the problem solved.

An additional remark concerns the selected mechanism of failure (Fig. 8.7). It is a relatively simple one. Despite its simplicity, it leads to quite involved algebraic calculations, especially when the micropile action is introduced (see the Appendix 8.2).

8.6 Mitigation Measures

Once the tunnel face becomes unstable the situation is shown in the sketch of Figure 8.21. A disordered soil-like material invades the tunnel hopefully creating a slope S (Fig. 8.21) which stabilizes the remaining debris inside the front cavern. In order to continue the excavation of the tunnel, the following sequence of actions could be adopted:

a) Avoid removing the debris inside the tunnel. They help to stabilize the face.
b) Some extra support may be gained by shotcreting (S_c) (Fig. 8.21) the exposed surface of the debris.
c) Build a bulkhead or wall on the front (W) (Fig. 8.21) in order to close the open cavity.
d) Fill the front cavern with a mortar or grout injected from the interior of the tunnel. A low pressure injection is suitable. This filling and injection process may be performed in several phases. The debris D (Fig. 8.21) should be also injected with the aim of creating an improved material.
e) Once the debris is stabilized and strengthened, boring operations could be resumed.

In the case of La Floresta tunnels, the face was protected by an umbrella of micropiles. This is an expensive procedure but it may be justified if applied in relatively short lengths.

Other possibilities (stabilizing buttress, face reinforcement) are sketched in Figure 8.8. If the soil or rock can be injected by means of cement grout or chemical resins, the soil/rock ahead of the face will be improved. These operations may be performed from the tunnel itself or from the surface if at all possible. Soil improvement ahead of the face, often combined with drainage, is a procedure used to cross faults.

Finally, the classical procedures to excavate soft ground, among them the so-called Belgian and German procedures (Szechy, 1967) are an alternative to improve stability conditions of the face. They may be combined with some of the reinforcement or improving soil techniques mentioned.

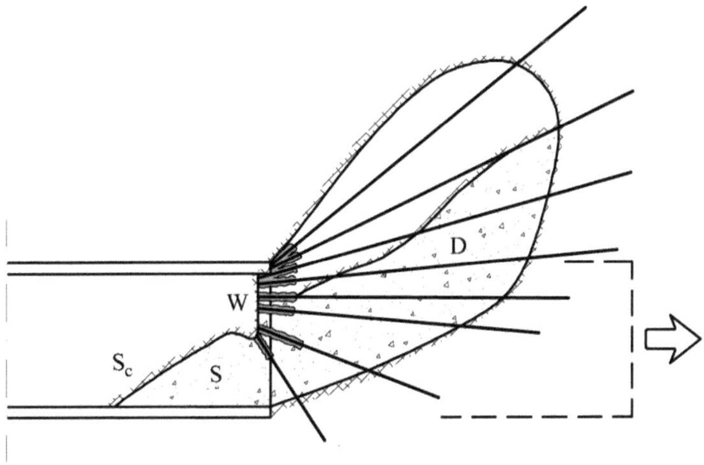

Figure 8.21 Sketch of tunnel face failure.

8.7 Lessons Learned

8.7.1 Face instability

Tunnel face instability is a risk associated with construction methods which do not provide face support (such as the new Austrian tunnelling method).

The face failures in the La Floresta tunnels were ultimately explained by the low strength of highly decomposed and densely fissured paleozoic shales which behaved as a soil with very low friction angle ("undrained" conditions have been used to interpret the failures).

8.7.2 Calculation procedures

At present, two-dimensional and three-dimensional stability solutions for face instability conditions are well developed and they can be used to estimate safety factors. The chapter describes a classical solution based on the application of the upper bound theorem of plasticity. In the case of La Floresta, strength values derived from the interpretation of face failures and a portal sliding slope on the same material were fairly consistent. This result adds confidence to the face stability analysis performed.

8.7.3 Stabilizing the face

Different procedures to stabilize the face of open face tunnels are available which have been mentioned in the chapter. Among them, the installation of micropile umbrellas is a popular procedure.

8.7.4 Stability of reinforced tunnel face

The upper bound theorem of plasticity offers the opportunity to analyze situations not properly addressed in published solutions of tunnel face instability. A design procedure has been described in the chapter. It is based on two aspects: 1) defining the limiting resisting conditions of the individual micropiles, and 2) including the micropile forces within the formulation of the upper bound theorem.

Micropile-limiting resisting forces have been calculated starting from the basic yield criterion for the steel reinforcement (Von Mises). Then, the upper bound plasticity solutions have been found for plane strain, undrained conditions and have been represented in graphs ready to use.

8.7.5 Finding safety factors

The safety factor associated with the reinforcement actually used in the stabilization of La Floresta tunnels has been estimated, on the basis of the developed procedure, by comparing the "in situ" undrained strength with the undrained strength necessary for equilibrium in the case of a reinforced tunnel face. This is probably the simplest way to define a safety factor in this case and it is consistent with the usual definition of safety factor in limiting equilibrium methods.

8.8 Advanced Topics

Bound theorems of plasticity provide a useful tool to determine collapse conditions of tunnel faces under undrained and drained conditions. Davis *et al.* (1980) determined bounds for the plane strain (unreinforced) case described in this chapter. Kimura and Mair (1981) examined the effect of the unlined length (P in Fig. 8.7a) by means of experiments in the centrifuge. Leca and Dormieux (1990) calculated the upper and lower bounds for the drained case (soil described by an effective cohesion and friction angle). They used rigid blocks of conical shape to describe failure mechanisms which are kinematically compatible. They expressed the solutions by means of a "bearing capacity" formula, in terms of coefficients which depend on friction angle and cover ratio. They also found that the upper bound solutions are closer to centrifuge results. Upper bound solutions may be improved if more flexible failure mechanisms are analyzed. Sloan and Assadi (1994) found an analytical solution for the upper bound, undrained, plane strain case when the failure mechanism was described by five degrees of freedom.

Later, procedures based on the numerical application of the bound theorems have been reported. Augarde *et al.* (2003) revisited the classical two-dimensional undrained case and found very accurate solutions. In these procedures the failure mechanism is part of the solution. The upper bound theorem was used by Leca *et al.* (1997) to study the effect of reinforcing tunnel headings by means of a dense distribution of bolts (Fig. 8.8b).

Another calculation approach is to use limit equilibrium techniques (Anagnostou and Kovari, 1996). They also provided their results in terms of "bearing capacity" expressions.

Finite element and distinct element methods have been used extensively to examine face stability, in most cases under three-dimensional conditions (Vermeer *et al.*, 2002; Ng and Lee, 2002; Galli *et al.*, 2003; Yoo and Shin 2003; Melis and Medina, 2005). Among them, Vermeer *et al.* (2002) determined failure conditions of the face by means of a "*c*, ϕ reduction method" and provided three-dimensional solutions for the drained case, ready to use, by means of plots and approximate analytical equations providing the stability factors of a "bearing capacity" formula. Sensitivity analyses described in most of the papers using numerical techniques provide a better understanding of the mechanism of face collapse and of the effect of dense bolt reinforcement at the tunnel heading.

Appendix 8.1 Von Mises' Yield Criterion

Von Mises' yield surface is defined by the following equation:

$$F(\boldsymbol{\sigma}) = \sqrt{3J_2'} - \sigma_e = 0, \tag{A8.1}$$

where σ_e is the elastic limit of the material and J_2' is the second invariant of the deviatoric stress tensor which can be expressed as follows:

$$\begin{bmatrix} \sigma_x - \sigma_m & \tau_{xy} & \tau_{xy} \\ \tau_{xy} & \sigma_y - \sigma_m & \tau_{yz} \\ \tau_{xz} & \tau_{yz} & \sigma_z - \sigma_m \end{bmatrix}, \tag{A8.2}$$

where σ_m is the mean stress, $\sigma_m = (\sigma_x + \sigma_y + \sigma_z)/3$. J_2' defines in a principal stress space ($\sigma_1, \sigma_2, \sigma_3$) the distance of a stress point to the hydrostatic axis and it can be related to the octaedric shear stress as

$$J_2' = \frac{3}{2}\tau_{oct} = \frac{\sqrt{3}}{2}\left(\sigma_1^2 + \sigma_2^2 + \sigma_3^2 - \frac{1}{3}\left(\sigma_1 + \sigma_2 + \sigma_3\right)^2\right). \tag{A8.3}$$

If Equation (A8.2) is introduced in (A8.1), Von Mises' yield surface is given by:

$$F(\boldsymbol{\sigma}) = \frac{1}{\sqrt{2}}\left(\left(\sigma_1 - \sigma_2\right)^2 + \left(\sigma_2 - \sigma_3\right)^2 + \left(\sigma_3 - \sigma_1\right)^2\right) - \sigma_e^2 = 0. \tag{A8.4}$$

Von Mises' criterion is only defined as a function of J_2'. Then, all the stress points of the yield surface are characterized by the same value of J_2'. The condition J_2' = constant defines a cylinder whose axis is the hydrostatic axis. Figure A8.1 shows the yield surface in the principal stresses space (Eq. (A8.4)).

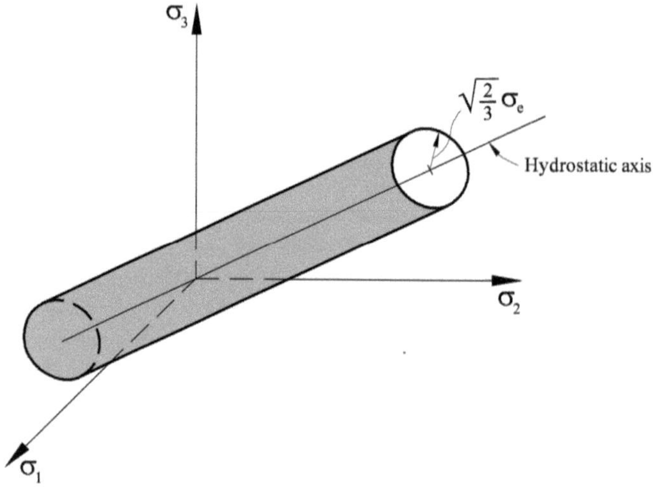

Figure A8.1 Von Mises Criterion

Consider now the stress point of a planar cross-section. This is the case of a beam subjected to normal and deviatoric forces as well as bending moment (Fig. A8.2). The stress tensor and the deviatoric stress tensor are given by:

$$\boldsymbol{\sigma} = \begin{bmatrix} \sigma_x & \tau_{xy} & 0 \\ \tau_{xy} & 0 & 0 \\ 0 & 0 & 0 \end{bmatrix}, \tag{A8.5}$$

$$\boldsymbol{\sigma}' = \begin{bmatrix} \frac{2}{3}\sigma_x & \tau_{xy} & 0 \\ \tau_{xy} & -\frac{1}{3}\sigma_x & 0 \\ 0 & 0 & -\frac{1}{3}\sigma_x \end{bmatrix}. \tag{A8.6}$$

Then, J_2' results in

$$J_2' = \frac{1}{3}{\sigma_x}^2 + {\tau_{xy}}^2 \tag{A8.7}$$

and finally, the Von Mises' yield surface (Eq. (A8.1)) can be expressed as

$$F(\boldsymbol{\sigma}) = \sqrt{\sigma_x^{\ 2} + 3\tau_{xy}^{\ 2}} - \sigma_e = 0. \tag{A8.8}$$

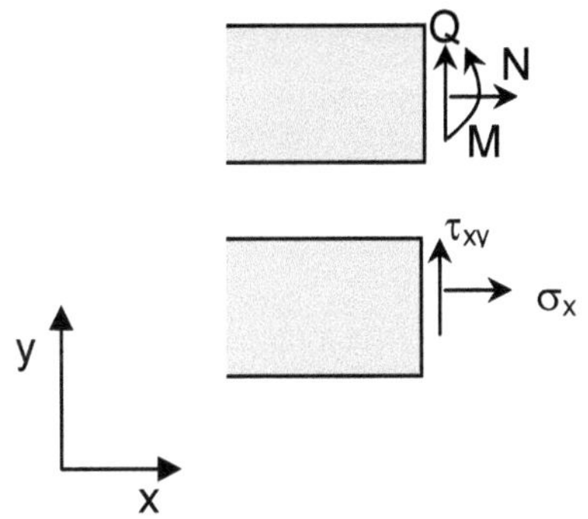

Figure A8.2 Forces and stresses on a planar cross-section of a beam.

Appendix 8.2 Upper Bound Solutions for Unreinforced and Reinforced Tunnel Faces

A8.2.1 Unreinforced face

This case is developed as an introduction to the more general case (reinforced face) considered in this chapter. The upper bound theorem of plastic collapse (Atkinson, 1981) states that

> *"If there is a set of external loads and a mechanism of plastic collapse such that the increment of work done by the external loads in an increment of displacement equals the work done by the internal stresses, collapse* ***must*** *occur and the external loads are an upper bound to the true collapse loads."*

The theorem is applied through the following steps:

- A kinematically compatible virtual failure mechanism is selected. The usual way is to divide the moving soil into a set of rigid bodies bounded by discontinuities. Work is only performed in discontinuities subjected to differential displacement across them.
- The work dissipated internally, due to a virtual deformation mechanism, is calculated. Under undrained conditions, the work dissipated per unit length of discontinuity is given by $c_u \cdot \delta_w$, where c_u is the undrained strength and δ_w is the virtual differential displacement.
- The work done by the external forces, when displaced by the assumed virtual failure mechanism, is calculated.
- External and internal works are made equal. The resulting equation allows the calculation of an expression for the external forces. They are a function of the geometrical parameters defining the mechanism.
- A minimization process, against the geometrical parameters defining the virtual mechanism, is then performed. Parameters defining the most

critical mechanism are calculated. These parameters, in turn, define the critical external force. This force is an upper bound to the actual failure load. Mechanisms closer to the actual ones and endowed with "high flexibility" (in terms of its capability to adopt a variety of geometrical configurations), lead to upper bounds closer to the actual failure loads.

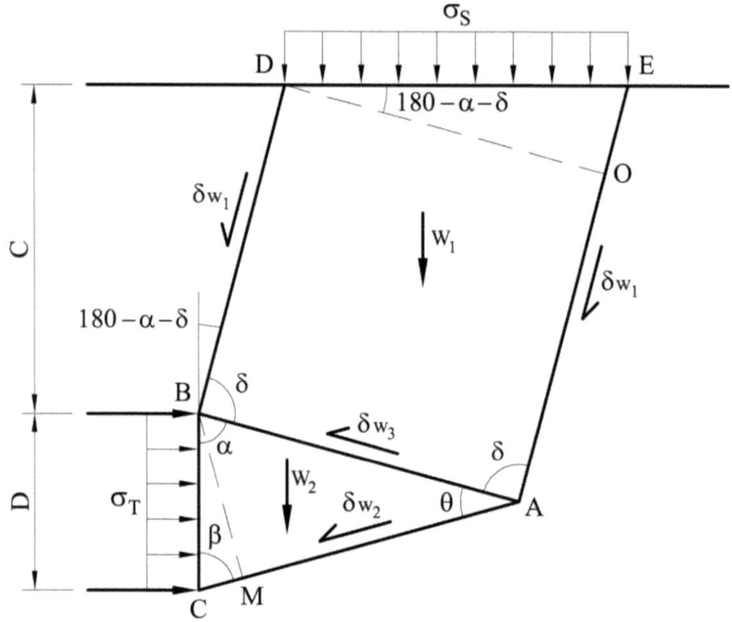

Figure A8.3 Failure mechanism for upper bound calculations: Unreinforced face. Plane strain.

Consider the plane strain geometry of the failure mechanism selected in Figure A8.3.

The geometry of the failure mechanism in Figure A8.3 is defined by lengths C (tunnel overburden depth) and D (tunnel diameter) and by the three angles (α, β, δ). In order to simplify the problem, the angle δ is assumed to be 90°. The external loads in the mechanism selected are defined by the stresses σ_S and σ_T applied to the soil surface and the tunnel face, respectively, and by the weight of the two wedges considered in the failure mechanism, W_1 and W_2. The relative displacements at the interfaces, δ_{W_1}, δ_{W_2} and δ_{W_3}, define the kinematics of the motion. They must satisfy a compatibility condition at point A, graphically expressed in Figure A8.4. The diagram results in

$$\delta_{W_2} = \frac{\delta_{W_1}}{\sin\theta}; \qquad \delta_{W_3} = \frac{\delta_{W_1}\cos\theta}{\sin\theta}. \tag{A8.9}$$

Note that $\theta = 180° - \alpha - \beta$.

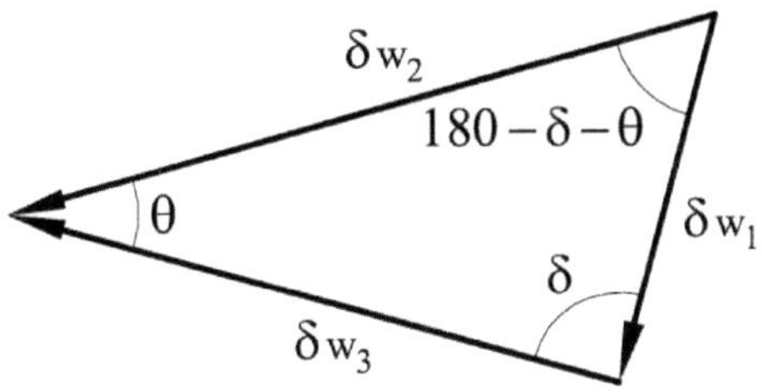

Figure A8.4 Displacement (velocity) diagram in point A.

The external work may be expressed (notation is given in Figure A8.3):

$$W_{\text{ext}} = \gamma \text{Area}_{\text{ABDE}} \delta_{w_1} \sin\alpha + \gamma \text{Area}_{\text{ABC}} \delta_{w_2} \cos\beta + \sigma_S L_{\text{DE}} \delta_{w_1} \sin\alpha - \sigma_T D \delta_{w_2} \sin\beta. \tag{A8.10}$$

The sign of the last term results from the opposite directions of the stress σ_T and the displacement δ_{w_2}.

The internal dissipation work is computed as

$$W_{\text{int}} = c_u \left[\left(L_{\text{BD}} + L_{\text{AE}} \right) \delta_{w_1} + L_{\text{AC}} \delta_{w_2} + L_{\text{AB}} \delta_{w_3} \right]. \tag{A8.11}$$

The following set of geometrical relationships helps to calculate the expressions for the external and internal work:

Lengths:

$$L_{\text{BD}} = \frac{C}{\sin\alpha}, \tag{A8.12a}$$

$$L_{\text{BM}} = D \sin\beta, \tag{A8.12b}$$

$$L_{\text{CM}} = D \cos\beta, \tag{A8.12c}$$

$$L_{\text{AB}} = L_{\text{DO}} = D \frac{\sin\beta}{\sin\theta}, \tag{A8.12d}$$

$$L_{\text{AM}} = D \frac{\sin\beta \cos\theta}{\sin\theta}, \tag{A8.12e}$$

$$L_{\text{AC}} = D \left(\frac{\sin\beta \cos\theta}{\sin\theta} + \cos\beta \right), \tag{A8.12f}$$

$$L_{\mathrm{DE}} = D\frac{\sin\beta}{\sin\theta\sin\alpha}, \tag{A8.12g}$$

$$L_{\mathrm{EO}} = D\frac{\sin\beta\cos\alpha}{\sin\theta\sin\alpha}, \tag{A8.12h}$$

$$L_{\mathrm{AE}} = D\frac{\sin\beta\cos\alpha}{\sin\theta\sin\alpha} + \frac{C}{\sin\alpha}. \tag{A8.12i}$$

Areas:

$$\begin{aligned} Area_{\mathrm{ABDE}} &= L_{\mathrm{AB}}L_{\mathrm{BD}} + \frac{L_{\mathrm{EO}}L_{\mathrm{DO}}}{2} = \\ &= DC\frac{\sin\beta}{\sin\theta\sin\alpha} + \frac{1}{2}D^2\frac{(\sin\beta)^2\cos\alpha}{(\sin\theta)^2\sin\alpha}, \end{aligned} \tag{A8.13}$$

$$Area_{\mathrm{ABC}} = \frac{1}{2}L_{\mathrm{AC}}L_{\mathrm{BM}} = D^2\sin\beta\left(\frac{\sin\beta\cos\theta}{\sin\theta} + \cos\beta\right). \tag{A8.14}$$

If expressions (A8.12), (A8.13) and (A8.14) are introduced in the work equations,

$$\begin{aligned} W_{\mathrm{int}} &= c_u\left[\left(L_{\mathrm{BD}} + L_{\mathrm{AE}}\right)\delta_{w_1} + L_{\mathrm{AC}}\delta_{w_2} + L_{\mathrm{AB}}\delta_{w_3}\right] = \\ &= c_u\delta_{w_1}\left[\frac{2C}{\sin\alpha} + D\left(\frac{\sin\beta\cos\alpha}{\sin\theta\sin\alpha} + \frac{2\sin\beta\cos\theta}{(\sin\theta)^2} + \frac{\cos\beta}{\sin\theta}\right)\right], \end{aligned} \tag{A8.15}$$

$$\begin{aligned} W_{\mathrm{ext}} &= \gamma Area_{\mathrm{ABDE}}\delta_{w_1}\sin\alpha + \gamma Area_{\mathrm{ABC}}\delta_{w_2}\cos\beta + \sigma_S L_{\mathrm{DE}}\delta_{w_1}\sin\alpha - \sigma_T D\delta_{w_2}\sin\beta = \\ &= \delta_{w_1}\gamma\left(DC\frac{\sin\beta}{\sin\theta} + \frac{1}{2}D^2\frac{(\sin\beta)^2\cos\alpha}{(\sin\theta)^2}\right) + \delta_{w_1}\gamma D^2\sin\beta\left(\frac{\sin\beta\cos\theta}{\sin\theta} + \cos\beta\right)\frac{\cos\beta}{\sin\theta} + \\ &\quad \delta_{w_1}\sigma_S D\frac{\sin\beta}{\sin\theta} - \delta_{w_1}\sigma_T D\frac{\sin\beta}{\sin\theta} = \\ &= \delta_{w_1}\left[\gamma\left(DC\frac{\sin\beta}{\sin\theta} + \frac{1}{2}D^2\frac{(\sin\beta)^2\cos\alpha}{(\sin\theta)^2}\right) + \gamma D^2\frac{\sin\beta\cos\beta}{\sin\theta}\left(\frac{\sin\beta\cos\theta}{\sin\theta} + \cos\beta\right) + \right. \\ &\quad \left.(\sigma_S - \sigma_T)D\frac{\sin\beta}{\sin\theta}\right]. \end{aligned} \tag{A8.16}$$

Making Equation (A8.15) equal to Equation (A8.16),

$$c_u\left[\frac{2C}{\sin\alpha}+D\left(\frac{\sin\beta\cos\alpha}{\sin\theta\sin\alpha}+\frac{2\sin\beta\cos\theta}{(\sin\theta)^2}+\frac{\cos\beta}{\sin\theta}\right)\right]=$$

$$=\gamma\left(DC\frac{\sin\beta}{\sin\theta}+\frac{1}{2}D^2\frac{(\sin\beta)^2\cos\alpha}{(\sin\theta)^2}\right)+ \tag{A8.17}$$

$$\gamma D^2\frac{\sin\beta\cos\beta}{\sin\theta}\left(\frac{\sin\beta\cos\theta}{\sin\theta}+\cos\beta\right)+(\sigma_S-\sigma_T)D\frac{\sin\beta}{\sin\theta}.$$

If $\sigma_S \geq \sigma_T$, Equation (A8.17) provides an upper bound for $(\sigma_S-\sigma_T)$ as a function of *C*, *D*, c_u and the angles α and β. It is convenient to introduce the following set of dimensionless variables: C/D, $\gamma D/c_u$ and $(\sigma_S-\sigma_T)/c_u$. Then, the upper bound theorem implies that

$$\frac{(\sigma_S-\sigma_T)}{c_u}\geq\frac{2C\sin\theta}{D\sin\alpha\sin\beta}+\left(\frac{\cos\alpha}{\sin\alpha}+\frac{2\cos\theta}{\sin\theta}+\frac{\cos\beta}{\sin\beta}\right)-$$

$$\frac{\gamma}{c_u}\left(C+\frac{1}{2}D\frac{\sin\beta\cos\alpha}{\sin\theta}\right)-\frac{\gamma D}{c_u}\cos\beta\left(\frac{\sin\beta\cos\theta}{\sin\theta}+\cos\beta\right). \tag{A8.18}$$

The minimization of (A8.18) with respect to α and β provides the minimum of the upper bound values of $(\sigma_s-\sigma_T)/c_u$ for the collapse mechanism defined. A further simplification is now introduced, assuming a more restricted family of mechanisms given by the condition $\alpha=\beta$. After some simple algebra, the following expressions for the external and internal work are obtained:

$$W_{\text{int}}=c_u\delta_{w_1}\frac{D}{\sin\alpha}\left[\frac{2C}{D}+\frac{1}{2}+\frac{(\sin\alpha)^2}{2(\cos\alpha)^2}\right], \tag{A8.19}$$

$$W_{\text{ext}}=\delta_{w_1}\frac{D}{2\cos\alpha}\left[\gamma D\left(\frac{C}{D}+\frac{1}{2}\right)+(\sigma_S-\sigma_T)\right]. \tag{A8.20}$$

Making equal the two work equations, the following upper bound is found:

$$\frac{(\sigma_S-\sigma_T)}{c_u}\leq\left(\frac{4C}{D}+1\right)\frac{1}{\tan\alpha}+\tan\alpha-\frac{\gamma}{c_u}D\left(\frac{C}{D}+\frac{1}{2}\right). \tag{A8.21}$$

In order to find the minimum of (A8.21), its derivative with respect to α is made equal to zero:

$$\frac{\partial}{\partial\alpha}\left[\frac{(\sigma_S-\sigma_T)}{c_u}\right]=0\ \Rightarrow$$

$$\Rightarrow\ \frac{\partial}{\partial\alpha}\left[\frac{4C}{D\tan\alpha}+\frac{1}{\tan\alpha}+\tan\alpha-\frac{\gamma D}{c_u}\left(\frac{C}{D}+\frac{1}{2}\right)\right]=0\ \Rightarrow \tag{A8.22}$$

$$\Rightarrow\ \tan\alpha=\sqrt{1+\frac{4C}{D}}=2\sqrt{\frac{1}{4}+\frac{C}{D}},$$

and substituting this value in (A8.21), the upper bound is found as follows:

$$\frac{(\sigma_S-\sigma_T)}{c_u}\le 4\sqrt{\frac{1}{4}+\frac{C}{D}}-\frac{\gamma D}{c_u}\left(\frac{C}{D}+\frac{1}{2}\right). \tag{A8.23}$$

It can be shown, at the cost of more cumbersome algebraic manipulations, that this solution is also the upper bound for a more general family of mechanisms defined for arbitrary values of (α,β,δ).

A8.2.2 Upper bound solution. Reinforced face

Consider now the case of a reinforced face as sketched in Figure A8.5. A set of steel tubular beams of length $L = a+b$, thickness t, spaced a distance s, inclined at an angle η with respect to the horizontal, cross the upper wedge of the failure mechanism and help to stabilize the tunnel front. The family of mechanisms considered now is geometrically identical to the mechanisms used in the analysis of the unreinforced face. In order to find an upper bound solution, the steps already described for the classical upper bound analysis will be followed. The displacement compatibility relations (A8.9) and the internal dissipation work (A8.15) will not change.

However, the external work will be modified. The virtual displacement rate δ_{w_1} (Fig. A8.5) results in a tensile (N_P) and a shear (Q_P) force which will oppose the motion of the wedge ABDE. Therefore, the external work will be now written

$$W_{\text{ext}}=\left(W_{\text{ext}}\right)_{\text{no\_beam}}+N_P\delta_{w_1}\cos\mu+Q_P\delta_{w_1}\sin\mu. \tag{A8.24}$$

The angle μ defines the direction of the virtual displacement with respect to the direction of the beam (see Fig. 8.9):

$$\mu=\alpha-\eta. \tag{A8.25}$$

The maximum forces developed by the reinforcement will take the steel to yielding conditions. It has been assumed that the yield condition is reached when the first fiber of the steel section yields. It has already been mentioned in section 8.5 that a fully plastified steel section may reach somewhat higher resisting forces.

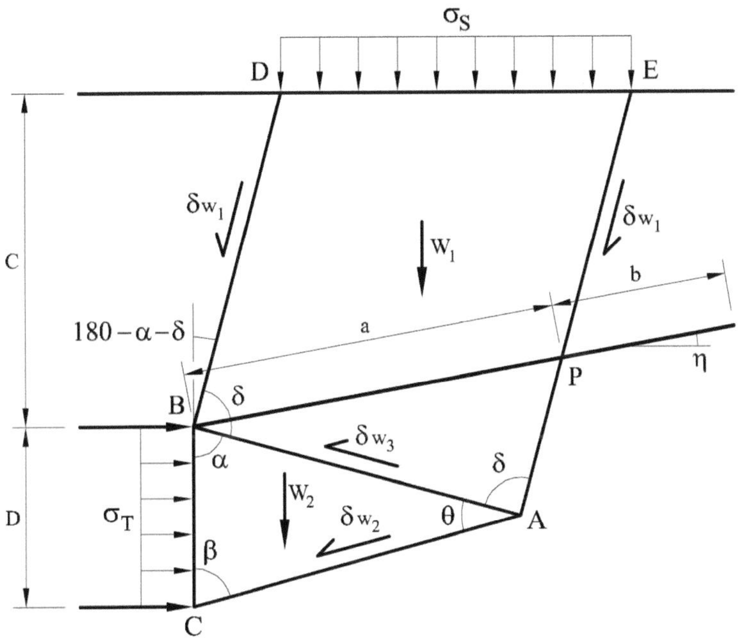

Figure A8.5 Failure mechanism for upper bound calculations: Reinforced face. Plane strain.

The assumption is now that once this situation is reached, the resisting forces will remain constant for increasing relative displacements. This hypothesis is justified by the ductile properties of reinforcement steel and its perfectly plastic behaviour. Expressions for N_P and Q_P were already found (Eqs. (8.15) and (8.16)). Expanding Equation (A8.24) results in

$$W_{\text{ext}} = \delta_{w_1}\frac{sD}{2\cos\alpha}\left[\gamma D\left(\frac{C}{D}+\frac{1}{2}\right)+\left(\sigma_s-\sigma_T\right)\right]+N_P\delta_{w_1}\cos\mu+Q_P\delta_{w_1}\sin\mu = \tag{A8.26}$$

$$= \delta_{w_1}\frac{sD}{2\cos\alpha}\left[\gamma D\left(\frac{C}{D}+\frac{1}{2}\right)+\left(\sigma_s-\sigma_T\right)+\frac{2\cos\alpha}{sD}N_P\cos\mu+\frac{2\cos\alpha}{sD}Q_P\sin\mu\right],$$

Making it equal to the internal dissipation work (Eq. (A8.15)), the expression (A8.27) can be obtained:

$$\frac{\left(\sigma_S-\sigma_T\right)}{c_u}=\frac{1}{\tan\alpha}\left(\frac{4C}{D}+1\right)+\tan\alpha-\frac{\gamma D}{c_u}\left(\frac{C}{D}+\frac{1}{2}\right)+\frac{2\cos\alpha}{c_u sD}N_P\cos\mu+\frac{2\cos\alpha}{c_u sD}Q_P. \tag{A8.27}$$

If expressions for the tensile normal and shear forces (Eqs. (8.15) and (8.16)) are introduced in Equation (A8.27):

$$\frac{(\sigma_S-\sigma_T)}{c_u}=\frac{1}{\tan\alpha}\left(\frac{4C}{D}+1\right)+\tan\alpha-\frac{\gamma D}{c_u}\left(\frac{C}{D}+\frac{1}{2}\right)+ \frac{\sigma_e td}{c_u sD}\frac{2\pi\cos\alpha}{\sqrt{f\left(\frac{d}{b},\mu\right)}}\left(\cos^2\mu+\frac{3}{2}\left(\frac{d}{b}\right)^2\sin^2\mu\right). \tag{A8.28}$$

Then, $(\sigma_S-\sigma_T)/c_u$ can be expressed in terms of dimensionless parameters and the angle α which define the geometry of the collapse mechanism:

$$\frac{(\sigma_S-\sigma_T)}{c_u}=F\left(\frac{C}{D},\frac{\gamma D}{c_u},\frac{\sigma_e td}{c_u sD},\frac{d}{b},\alpha\right). \tag{A8.29}$$

Note that Equation (A8.25) has been considered in (A8.29).

The dimensionless parameter, $\sigma_e td/c_u sD$, is an expression that quantifies the strength provided by the micropile with respect to the strength provided by the soil.

The minimum upper bound of $(\sigma_S-\sigma_T)/c_u$ requires the minimization of (A8.29) with respect to angle α. The derivative of (A8.29) was formally obtained with the help of the Maple program. It is a long and cumbersome expression which is not included here. However, no analytical solutions for the critical α (which minimizes the function F (Eq. (A8.29)) and the root of its derivative ($\partial F/\partial\alpha$)) could be found.

Different numerical procedures can be used in order to obtain the solution. In this case, the solver function available in the Excel Program has been used. It allows finding the root of a function and also its maximum and minimum values. The results have been plotted in Figures 8.12 to 8.19.

References

Anagnostou, G. and Kovàri, K. (1996) Face stability conditions with earth-pressure-balanced shields. *Tunnelling and Underground Space Technology* 11 (2), 165 – 173.

Atkinson, J.H. (1981) *Foundations and Slopes. An Introduction to Applications of Critical State Soil Mechanics.* McGraw Hill, London.

Augarde, C.E., Andrei, V.L. and Sloan, S.W. (2003) Stability of an undrained plane strain heading revisted. *Computers and Geotechnics* 30, 419 – 430.

Bieniawski, Z.T. (1989) *Engineering Rock Mass Classifications.* Wiley, New York.

Chen W.F. and Liu, X.L. (1990) *Limit Analysis in Soil Mechanics.* Elsevier, Amsterdam.

Davis, E.H., Gunn, M.J., Mair, R.J. and Seneviratne, H.N. (1980) The stability of shallow tunnels and underground openings in cohesive material. *Géotechnique* 30 (4), 397 – 416.

Ewing, D.J.F. and Hill, R. (1967) The plastic constraint of V-notched tension bars. *Journal of the Mechanics and Physics of Solids* 15, 115 – 124.

Galli, G., Grimaldi, A. and Leonardi, A. (2003) Three-dimensional modelling of tunnel excavation and lining. *Computers and Geotechnics* 31, 171 – 183

Kimura, T. and Mair, R.J. (1981) Centrifugal testing of model tunnels in soft clay. *Proceedings of the 10th International Conference Soil Mechanics and Foundation Engineering*, Stockholm, 1, 319 – 222.

Leca, E. and Dormieux, L. (1990) Upper and lower bound solutions for the face stability of shallow circular tunnels in frictional material. *Géotechnique* 40 (4), 581 – 606.

Leca, E., Garnier, J., Atwa, M., Chambon, P., Skiker, A., Dormieux, L., Garnieer, D. and Maghous, S. (1997) Analyse théorique et expérimentale de la stabilité du front de taille des tunnels à faible profondeur. *Proceedings of the 14th International Congress on Soil Mechanics and Foundation Engineering*, Hambourg, 3, 1421 – 1424.

Lee, I.M., Lee, J.S. and Nam, S.W. (2004) Effect of seepage force on tunnel face stability reinforced with multi-step pipe grouting. *Tunnelling and Underground Space Technology* 19, 551–565.

Melis, M.J. and Medina, L.E. (2005) Discrete numerical model for analysis of earth pressure balance tunnel excavation. *Journal of Geotechnical and Geoenvironmental Engineering* 131 (10), 1234 – 1242.

Ng C.W.W. and Lee G.T.K. (2002) A three-dimensional parametric study of the use of soil nails for stabilising tunnel faces. *Computers and Geotechnics* 29, 673–697.

Poulos, H.G. and Davis, E.H. (1980) *Pile Foundation Analysis and Design*. Wiley, New York.

Sloan, S.W. and Assadi, A. (1994) Undrained stability of a plane strain heading. *Canadian Geotechnical Journal* 31, 443 – 50.

Szechy, K. (1967) *The Art of Tunneling*. Akdemiai Kiado, Budapest.

Terzaghi, K. (1955) Evaluation of coefficients of subgrade reaction. *Géotechnique* 4, 297-326.

Vermeer, A., Ruse, N. and Marcher, T. (2002) Tunnel Heading Stability in Drained Ground. *Felsbau* 20 (6), 1 – 17.

Yoo, C. and Shin, Y.K. (2003) Deformation behaviour of tunnel face reinforced with longitudinal pipes—laboratory and numerical investigation. *Tunnelling and Underground Space Technology* 18, 303 – 319.

Young, W.C. (1989) *Roark's Formulas for Stress and Strains*. McGraw Hill, New York.

Epilogue

If the reader had patience to make it to this point, the main objective of this book has, probably, been achieved. It is not an easy task to fascinate a student with a standard course on Soil Mechanics and Geotechnical Engineering. If, however, the same material is presented as a tool to explore a natural or a man-made "disaster", both the motivation and the ability to absorb this material increase dramatically. Using this approach in the educational process has proven to be extremely beneficial and can be highly recommended. The case studies in this book could help to build an introductory course, covering such basic topics as settlements, bearing capacity and excavations.

The failure cases considered in this book have one thing in common – they can be all reasonably well explained using the so called "back-of-the-envelope" calculations, i.e., without sophisticated models requiring finite element analysis. These simple methods, based on clear mechanical considerations, are the endangered species of the computer dominated era, and should, in the authors' opinion, be protected from extinction. And not just for nostalgic reasons – they could sometimes prevent a disaster caused by a wrong application of computer models (e.g., Nicoll Highway). In particular, the upper bound limit analysis has proven itself again as a powerful tool, allowing for the sufficiently accurate estimate of the failure loads and leaving a lot of room for creativity (e.g., in non-homogeneous soils, like in Transcona Grain Elevator and Barcelona Port Caissons failures).

Nobody is exempt from making mistakes. But repeating the mistakes, which are well known within the profession, reveals a gap in education. Therefore, another objective of this book has been to make one more step towards closing this gap. Of course, many more failure cases covering a much larger area of geotechnical problems can be considered, some of them being included into the companion book *Geomechanics of Failures. Advanced Topics* by the same authors. The present book, however, is meant to give a taste of learning the soil mechanics from "other people's mistakes", and, hopefully, to raise the appetite for wanting to learn more. The authors, in their turn, have found investigating failure cases to be a truly eye opening experience, and see this book as a work in progress, which will be, hopefully, reflected in future editions.

MIX
Papier aus verantwortungsvollen Quellen
Paper from responsible sources
FSC® C105338

If you have any concerns about our products,
you can contact us on
ProductSafety@springernature.com

In case Publisher is established outside the EU,
the EU authorized representative is:
Springer Nature Customer Service Center GmbH
Europaplatz 3, 69115 Heidelberg, Germany

Printed by Libri Plureos GmbH
in Hamburg, Germany